Lecture Notes in Computer Scien<

T0238359

Commenced Publication in 1973
Founding and Former Series Editors:
Gerhard Goos, Juris Hartmanis, and Jan van Leeuwen

Zohra Bellahsène   Ela Hunt
Michael Rys   Rainer Unland (Eds.)

Database and XML Technologies

6th International
XML Database Symposium, XSym 2009
Lyon, France, August 24, 2009
Proceedings

 Springer

Volume Editors

Zohra Bellahsène
LIRMM - UMR 5506 CNRS, Université Montpellier II
Montpellier, France
E-mail: bella@lirmm.fr

Ela Hunt
Department of Computer Science, ETH Zurich
Zurich, Switzerland
E-mail: elahunt@inf.ethz.ch

Michael Rys
Microsoft Corporation
Redmond, WA, USA
E-mail: mrys@microsoft.com

Rainer Unland
ICB, University of Duisburg-Essen
Essen, Germany
E-mail: unlandr@informatik.uni-essen.de

Library of Congress Control Number: 2009932605

CR Subject Classification (1998): H.2, H.3, H.4, H.2.3, H.2.4, H.3.3, H.2.8

LNCS Sublibrary: SL 3 – Information Systems and Application, incl. Internet/Web
and HCI

ISSN   0302-9743
ISBN-10   3-642-03554-X Springer Berlin Heidelberg New York
ISBN-13   978-3-642-03554-8 Springer Berlin Heidelberg New York

springer.com

© Springer-Verlag Berlin Heidelberg 2009
Printed in Germany

Typesetting: Camera-ready by author, data conversion by Scientific Publishing Services, Chennai, India
Printed on acid-free paper   SPIN: 12728664   06/3180   5 4 3 2 1 0

Preface

Since its first edition in 2003, the XML Database Symposium series (XSym) has been a forum for academics, practitioners, users and vendors, allowing all to discuss the use of and synergy between database management systems and XML. The previous symposia have provided opportunities for timely discussions on a broad range of topics pertaining to the theory and practice of XML data management and its applications. XSym 2009 continued this XSym tradition with a program consisting of 15 papers and a keynote shared with the 12th International Symposium on Database Programming Languages (DBPL 2009). We received 26 paper submissions, out of which eight papers were accepted as full papers, and seven as short/demo papers. Each submitted paper underwent a rigorous and careful review by four referees for long papers and three for the short ones.

The contributions in these proceedings are a fine sample of the very best current research in XML query processing, including full text, keyword and loosely structured queries, stream querying and joins, and materialized views. Among new theoretical advances we included work on a lambda-calculus model of XML and XPath, on mapping from the enhanced entity-relationship conceptual model to the W3C XML Schema Language, on transactions, and extensions to XPath. Finally, work on data parallel algorithms, compression, and practical aspects of XQuery, including query forms and the use of Prolog are also part of this volume.

The organizers would like to express their gratitude to the authors, for submitting their work, and to the Program Committee, for providing very thorough evaluations of the submitted papers and for the discussions that followed under significant time constraints. We also would like to thank the invited keynote speaker for the challenging and thought-provoking contribution. Finally, we are also grateful to Microsoft for their generous sponsorship, Andrei Voronkov and other contributors for the EasyChair conference management system, and the local organizers for their effort in making XSym 2009 a pleasant and successful event. Finally, we would also like to thank Alfred Hofmann and his great team from Springer for their support and cooperation in putting this volume together.

June 2009

Zohra Bellahsène
Ela Hunt
Michael Rys
Rainer Unland

Organization

Steering Committee

Zohra Bellahsène	LIRMM-CNRS/University Montpellier 2, France
Ela Hunt	University of Strathclyde, Scotland, UK
Michael Rys	Microsoft, USA
Rainer Unland	University of Duisburg-Essen, Germany

Local Organization Chairs

Mohand Said Hacid	Université Lyon 1, France
Jean Marc Petit	INSA Lyon, France

International Program Committee

Alfredo Cuzzocrea	CNR, Italy
Angela Bonifati	CNR, Italy
Ashraf Aboulnaga	University of Waterloo, Canada
Bernd Amann	LIP 6 Université Pierre et Marie Curie, France
Carl-Christian Kanne	University of Mannheim, Germany
Denilson Barbosa	University of Alberta, Canada
Elisabeth Murisasco	Université du Sud Toulon, France
Emmanuel Waller	LRI, France
Giovanna Guerrini	Università di Genova, Italy
Hakim Hacid	Université Lyon 2, France
Ingo Frommholz	University of Glasgow, Scotland, UK
Ioana Manolescu	INRIA, France
Irini Fundulaki	ICS-Forth, Greece
Jeffrey Xu Yu	University of Hong Kong, China
John Wilson	University of Strathclyde, Scotland, UK
Luc Moreau	University of Southampton, UK
Marc Scholl	Universität Konstanz, Germany
Marco Mesiti	Università di Milano, Italy
Matthias Nicola	IBM, USA
Mirian Halfeld Ferrari Alves	Université François Rabelais de Tours, France
Nilesh Dalvi	Yahoo, USA
Nikolaus Augsten	Free University of Bozen-Bolzano, Italy
Norman May	SAP, Germany
P. Sreenivasa Kumar	IIT Madras, India
Peter Fischer	ETH Zurich, Switzerland
Peter McBrien	Imperial College, London, UK

Peter Wood	Birkbeck College, University of London, UK
Sourav S Bhowmick	NTU, Singapore
Stephane Bressan	NUS, Singapore
Stratis Viglas	University of Edinburgh, Scotland, UK
Tadeusz Pankowski	Adam Mickiewicz University, Poznan, Poland
Werner Nutt	Free University of Bozen-Bolzano, Italy
Zografoula Vagena	Microsoft, UK

External Reviewers

| Christian Grün | Universität Konstanz, Germany |

Table of Contents

XML Transaction Management and Schema Design

Ordered Backward XPath Axis Processing against XML Streams

Abdul Nizar M. and P. Sreenivasa Kumar

Dept. of Computer Science and Engineering
Indian Institute of Technology Madras, Chennai, 600 036, India
{nizar,psk}@cse.iitm.ac.in

Abstract. Processing of backward XPath axes against XML streams is challenging for two reasons: (i) Data is not cached for future access. (ii) Query contains steps specifying navigation to the data that already passed by. While there are some attempts to process *parent* and *ancestor* axes, there are very few proposals to process ordered backward axes namely, *preceding* and *preceding-sibling*. For ordered backward axis processing, the algorithm, in addition to overcoming the limitations on data availability, has to take care of ordering constraints imposed by these axes. In this paper, we show how backward ordered axes can be effectively represented using forward constraints. We then discuss an algorithm for XML stream processing of XPath expressions containing ordered backward axes. The algorithm uses a layered cache structure to systematically accumulate query results. Our experiments show that the new algorithm gains remarkable speed up over the existing algorithm without compromising on bufferspace requirement.

1   Introduction

Due to the wide-spread use of XML data, especially in the web context, content- and structure-based filtering and information extraction from streamed XML documents attracted interest of the research community. Majority of proposals that process XPath expressions consisting of *child* ('/') and *descendant* ('//') axes against XML streams use twigs to encode the query expression.

There are very few attempts to tackle the more challenging task of processing backward (*parent, ancestor, preceding, preceding-sibling*) axes against XML streams. The $\chi\alpha o\varsigma$[1] system processes *parent* and *ancestor* axes in addition to *child* and *descendant* axes while the SPEX system[2] processes all backward and forward axes. In this paper, we present an efficient stream querying algorithm for XPath expressions with ordered backward axes –*preceding* and *preceding-sibling* axes in addition to *child, descendant* axes. Ordered forward axis processing is discussed elsewhere [3].

The contributions of this paper are: (1) We show how conventional twig structure can be extended to represent the semantics of ordered backward axes – *preceding* and *preceding-sibling* – using forward constraints. (2) We present a stream querying algorithm for XPath expressions with those axes in addition

Z. Bellahsène et al. (Eds.): XSym 2009, LNCS 5679, pp. 1–16, 2009.

to *child*, *descendant* axes. (3) We experimentally show that the new algorithm gains remarkable speed up over the existing algorithm without compromising on bufferspace requirement.

The rest of the paper is organized as follows: Section 2 presents the background, motivates the work and glances over the related literature. In Section 3 we discuss the new framework and the matching algorithm. Section 4 presents experimental results and Section 5 concludes the paper.

2   Motivation and Related Work

Stream processing of ordered backward axes, namely *preceding* and *preceding-sibling*, is challenging for three reasons: (i) The streaming model assumes that data is not cached for future access. (ii) The query contains steps specifying navigations to data which have already passed by. (iii) The algorithm has to take care of ordering constraints dictated by ordered axes.

A twig is a tree where nodes are labelled with node tests and edges are directed. It is used to represent XPath expressions with *child* and *descendant* axes (see Fig. 1; twigs T_{1a} and T_{1b} are isomorphic and equivalent). In a twig, edges labelled '//' (A-D edges) represent *descendant* axes while the remaining edges (P-C edges) represent *child* axes. The node labelled r is the twig root and is used to distinguish between twigs representing absolute (those starting with '/') and and relative (those starting with '//') path expressions. The *result* node is shown in black.

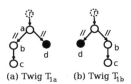

(a) Twig T_{1a} (b) Twig T_{1b}

Fig. 1. Twigs Representing /a[.//b/c]//d

There have been many proposals to process un-ordered XPath axes (*child, descendant, parent, ancestor*) against streaming data. Majority of these algorithms use the basic twig structure and its extensions to encode the query expression and systematically find matches [4,5,6,7]. The efficiency of twig-based algorithms is attributed to the fact that completely nested structure of XML document can be effectively combined with LIFO nature of stacks attached to the twig nodes to find matches.

There are queries that can be effectively expressed using ordered backward axes.

Example 1. In the context of surgical procedures which consist of a number of procedures, actions, observations etc., the order in which events take place is important. For instance the query "The observations that led to an immediate surgery" can be expressed as //action[.= "immediate surgical intervention"] /preceding::observation.

As the twig structure lacks order information, it can not convey the semantics of ordered axes and hence the algorithms based on twigs can not be directly extended to handle ordered axes. These circumstances motivate us to look into a new way of representing backward ordered axes and a new stream querying strategy based on the new model.

Related Work: XML stream querying is an active area of research and many systems have been proposed in recent past ([8],[6],[4],[7], [1],[9],[10]).

The $\chi\alpha o\varsigma$ system proposed by Josifovsky *et.al.*[1] performs stream processing of XPath expressions with *child, descendant, parent* and *ancestor* axes. The *TurboXPath* system[9] handles stream processing of XQuery-like queries. The XSQ system proposed in [10] handles complex XPath Queries with *child, descendant* and closure axes with predicates and can handle aggregations. The system uses a hierarchy of push-down transducers with associated buffers.

The TwigM algorithm proposed in [6] avoids proliferation of exponential number of twig matches in recursive XML streams by using compact encoding of potential matches using a stack structure. The system proposed by Aneesh *et.al.*[4] performs shared processing of twigs in document order by breaking twigs into branch sequences, which are sequences around branch points. The system due to Gou and Chirkova[7], which processes conventional twigs, achieves better time and space performance than TwigM when processing recursive XML documents. It can also handle predicates with boolean connectives. The authors propose two variants of the algorithm – Lazy Querying (LQ) and Eager Querying (EQ) – of comparable performance.

Ordered backward axis processing, to the best of our knowledge, is limited to the SPEX system[2]. The system processes an XPath expression by mapping it to a network of transducers. Most transducers used are single-state push-down automata with output tape. Each transducer in the network processes the XML stream it receives and transmits it, either unchanged or annotated with conditions, to its successor transducers. The transducer of the result node holds potential answers, to be output when conditions specified by the query are found to be true by the corresponding transducers.

The algorithm discussed in [3] handles forward axes. The approach of rewriting the expressions with backward axes into expressions using forward axes alone and using the solution in [3] does not work as the rewriting process introduces predicates with forward axes.

3   Handling Ordered Backward Axes

In this section, we discuss the algorithm for matching path expressions with *child, descendant, preceding-sibling* and *preceding* axes. We start with matching of expressions without predicates. Predicate handling is discussed in Section 3.3.

3.1   Representing Ordered Axes

We extend conventional twigs to what we call *Order-aware Twigs(OaTs)* to represent XPath expressions with backward ordered axes. Note that the match

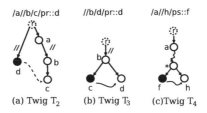

Fig. 2. Order-aware Twigs

of a twig against an XML document is conventionally defined as a mapping from the nodes in the twig to the nodes in the document satisfying twig-node labels and relationships between twig-nodes.

LR- and SLR-Orderings: To represent backward ordered axes, we add two types of forward ordering constraints – *LR (Left-to-Right)-Ordering* and *SLR (Sibling Left-to-Right)-Ordering* – to conventional twigs. LR-ordering is specified from a node x to node y such that x and y appear in two *disjoint* downward paths from the twig root r. It has the interpretation that, in a match of the twig against some document D, the nodes – say p and q – matching x and y should be such that q appears after p in document order in D, but is *not* a descendant of p. LR-Ordering can effectively represent *preceding* axis appearing in the path expressions. For instance, the expression /a//b/c/pr::d[1] looks for d-nodes which are appearing before the opening-tag of, but not an ancestor of, a c-node child of a b-node descendant of the document root node a. To represent the axis pr::d, a new node labelled d is connected to the root of the twig through an A-D edge and an LR-edge (dashed edge) is added from node d to c (see twig T_2 in Fig. 2(a)). Clearly, semantics of T_2 is the same as that of the query.

An SLR-Ordering is specified from a node x to a sibling node y such that x and y are connected to their parent via P-C edges. It has the interpretation that in a match of the twig against some document D, the nodes – say p and q – matching x and y should be such that p and q are siblings and q appears after p in document order in D. SLR-Ordering can effectively represent *preceding-sibling* axis. The OaT T_3 in Fig. 2(b) represents the XPath expression //b/d[ps::c] where SLR-Ordering is shown using a solid arrow from c to d. If there is an (S)LR-Ordering from node n_1 to node n_2 in an OaT, n_1 is called the *tail* of the (S)LR-Ordering and n_2 is called the *head*.

Closure Edges: The basic OaTs need to be further extended to handle XPath expressions containing an axis step with *preceding-sibling* axis that appears *immediately* after an axis step with *descendant* axis. For example, in the query /a//h/ps::f, f can be either left-sibling of an h-child of a or left-sibling of h-child of descendant of a. We handle this situation by introducing a new type of relationship edge known as the *closure edge*. A closure edge from node n_1 to a

---

[1] We use pr and ps as abbreviations for preceding and preceding-sibling, respectively.

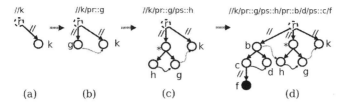

Fig. 3. Illustrating Transformation of Path Expression to OaT T_5

node n_2 with a wild card label ('*') indicates that in a match of the OaT, the document node d_2 matching with n_2 can be either the same as the document node d_1 matching with n_1 or a descendant of d_1. n_2 is called *closure node*. Twig T_4 in Fig. 2(c) shows the use of closure-edge to represent the path expression /a//h/ps::f. Here the *zig-zag* edge between the a-node and the closure node is a closure edge.

Figure 3 illustrates how an expression with *child, descendant, preceding-sibling* and *preceding* axes is systematically translated into the equivalent OaT. It illustrates a special case where the expression has a *preceding-sibling* axis appearing *immediately* after a *preceding* axis. A wild card node is introduced as parent of h and g and an A-D edge is added from r to the wild card node. Note that there is no need of a closure-edge from r to the dummy node as r is assumed to match to a cosmic root that sits above the root of a document tree. Please see [3] for a detailed account of translation of XPath expressions to OaTs.

From the discussion above and Fig. 3, it is clear that an axis step of the form *a/preceding-sibling::b* adds a new P-C edge from parent of *a* to a new node *b* and an SLR ordering from *b* to *a*. Similarly, *a/preceding::b* adds a new A-D edge in the OaT from r to a new node *b* and an LR-edge from *b* to *a*. We term the sub-tree rooted at a child of r a *branch-tree*.

If we assume that SLR and LR-Orderings are from left to right, the *deepest, left-most* node in the *left-most* branch-tree is the result node and the deepest, left-most node in each of the remaining branch-trees is an LR-head node. It may also be noted that an LR-tail node is *always* a child or grand child of r.

Node Types: We classify nodes in the OaT into *five* types – *result, SLR-head, LR-head, SLR-parent* and *ordinary* nodes. A node with SLR-Ordering specified among its children is termed an SLR-parent node. A node not belonging to the first four types is an ordinary node.

If a node e exists in the path from the SLR-parent node to the result or LR-head node such that e is connected to its child by an A-D edge, it is called the *frontier node* of the SLR-parent node. The child of r in the *right-most* branch tree is called the *output node* as the results are output at this node.

In the OaT of Fig. 3(d), f is the result node, d is an SLR-head node, h is an LR-head node, b is an SLR-parent node and c is an ordinary node. Also, c is the frontier node for b. k is the output node of the OaT.

3.2   Matching Algorithm

As the input is a stream, the matching algorithm has to find the document nodes that match with the OaT nodes in terms of labels and relationships and, in that process, has to accumulate the nodes that are potential results. The algorithm maintains a stack at every OaT node. Each stack frame represents an element in the stream that matches with the query node to which stack is associated. Let N_j be a node in the OaT and N_i be its parent. The topmost frame F_p in $N_i.stack$ that matches with a frame F_c in $N_j.stack$ based on the relationship constraint (P-C or A-D) specified between N_i and N_j is called the parent frame of F_c. A stack frame has the following fields: (i) *levelNo*: depth, in the document tree of the node represented by the frame. (ii)*frameNo*: A link to the parent frame.

The algorithm starts by pushing a dummy frame $\langle 0, null \rangle$ into the stack for node r. It then responds to events generated by a SAX parser. The global variable *gDepth* (initialized to 1) is maintained by the algorithm to keep track of depth of nodes in the XML document tree being streamed in.

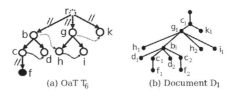

(a) OaT T_6 (b) Document D_1

Fig. 4. Sample Document and OaT

The challenge is to accumulate results respecting order constraints dictated by SLR- and LR-Orderings. For instance, in the context of OaT T_6 and document D_1 in Fig. 4, suppose, during query processing, the potential result node f_1 is accumulated in the frame for c_1 in c.*stack*. Since there is an SLR ordering from c to d, f_1 can be validated only after d_2 is seen. Note that f_2 accumulated in the frame for c_2 can not validated using d_2 as it appears *before* c_2.

LR-constraint handling is more involved. Once a potential result node satisfies the conditions specified on a branch tree, we can check if the node satisfies conditions laid down by the branch trees to the right. For instance, suppose that the potential result nodes accumulate at the frame for b_1 in b.*stack*. After processing the close-tag of b_1, these potential nodes satisfy conditions up to the OaT node b. Now we can check if the nodes satisfy conditions on the branch tree rooted at g. We use special kind of 'layered' caches associated with stack frames to effectively accumulate the results.

Layered Caches: An *s-cache* is associated with each stack frame. Number of layers in the s-cache is equal to the fanout of the OaT node to which the stack is associated and each layer is associated with one child of the node in left-to-right order. The stack frames in leaf nodes have single layer s-caches. Each stack frame of SLR-parent node has a layered *t-cache*. Number layers in this cache is one less than fanout of the node the layers are associated with children of the node other than the left-most one in left-to-right order. Each t-cache and s-cache layer of the stack frames for nodes in the left-most branch tree can hold a set of document node ids. In the case of stack frames of nodes in other branch-trees, the s-cache and t-cache layers hold bucket pointers.

The stack frame of r has an *rs-cache* and an *rt-cache* in addition to the s-cache. The rs-cache is similar to s-cache. Each rt-cache layer is a sequence of buckets with each bucket holding a set of document node ids. Number of rt-cache layers is one less than the number of branches of r and each layer is associated with one branch tree of r other than the left-most one in left-to-right order.

Processing *preceding-sibling* Axis: We start with the matching algorithm for path expressions without *preceding* axis and subsequently show how the algorithm can incorporate *preceding* axis processing. Due to space constraints, we exclude the discussion on closure node processing, the details of which are available in [3]. Note that, if the path expression has no *preceding* axis, the corresponding OaT will appear as a single branch under the root. Fig. 5(a) shows the partial OaT T_{6a} having the left-most branch of OaT T_6 in Fig. 4(a).

Algorithms 1 and 2 show the steps in *open-tag* event processing and *close-tag* event processing, respectively. In the algorithms, the s-cache layer of the stack frame F of a query node N_j corresponding to a child node N_k is denoted as $N_j.F.s\text{-}cache[N_k]$. We also use the notation $N_j.F.s\text{-}cache[i]$ to denote the ith s-cache layer. Similar notation is used for t-cache.

The algorithms assume the following abstract functions:- (i) *labelMap(x)*: For an element with tag x, returns OaT nodes N_j such that $N_j.label = x$ or $N_j.label =$ '*'. (ii) *makeFrame(d, framePosn, f)*: Returns a new stack frame in which *depth* and *frameNo* are equal to d and *framePosn*, respectively. It also creates an empty s-cache of with size f (and a t-cache of size $f - 1$ for SLR-parent nodes). For leaf nodes of the OaT, a single layer s-cache is created (iii) *reclaim(F)*: Reclaims the memory allocated to the stack frame F. (iv) *parent()* and *SLR-tail()*: Return parent and SLR-tail of an OaT node, respectively. (v) Standard stack functions *pop()*, *push()*, *top()* and *isEmpty()*. Here *top()* returns position of the topmost stack frame.

The handlers operate on the elements of the document stream, which arrives in document order. Intuitively, for an open-tag $\langle x \rangle$ of a document node e, the *open-tag* handler pushes a frame into the stack of the OaT node with label x, say N_j, *if* there is a chain of frames in the stacks of the nodes in the path from r to N_j satisfying the P-C and A-D relationships specified along the path. This condition can be checked incrementally by comparing depth of e with depth value of the *top-most* stack frame in the parent of N_j (Algorithm 1, lines 5–8). If the condition is satisfied, the top-most frame becomes parent frame of the frame for e. When the close-tag $\langle /x \rangle$ of e is seen, the close-tag handler pops out and processes the corresponding frame (Algorithm 2, lines 4–6). The actual steps vary depending on the node type of N_j.

Algorithm 1. Open-Tag Event Handler

Data: Open tag $\langle x \rangle$ of an element in the stream

1 **Open-Tag-Handler**(x)

2 $d \leftarrow gDepth \leftarrow gDepth + 1$;

3 **foreach** $N_j \in$ **labelMap**(x) **do**

4    $N_p \leftarrow parent(N_j)$

5   **if** **isEmpty**$(N_p.Stack)$ **then**

6      $\lfloor$ Continue with the next iteration;

7   **if** $(N_j$ is a child node$) \wedge (d - N_p.Stack[top(N_p)].levelNo > 1)$ **then**

8      $\lfloor$ Continue with the next iteration;

9    $F \leftarrow$ **makeFrame**$(d, top(N_i.stack), N_j.fanOut)$

10   **switch** $(N_j.nodetype)$ **do**

11   **case** result node

12       $\lfloor$ Add e_{id} to $F.s\text{-}cache[1]$;

13   **case** SLR-head node

14       $N_t \leftarrow$ SLR-tail(N_j);

15   **if** $N_p.stack[top(N_p)].s\text{-}cache[N_t]$ is empty **then**

16         $\lfloor$ Continue with the next iteration;

17   Move id tuples in $N_p.stack[top(N_p)].s\text{-}cache[N_t]$ to
         $N_p.stack[top(N_p)].t\text{-}cache[N_j]$;

18   $\lfloor$ **push**$(N_j.Stack, F)$

Open-Tag Handler: If N_j is an ordinary node or SLR-parent node, open-tag handler pushes a frame for e to $N_j.stack$ (Algorithm 1, line 18). If N_j is a result node, the open-tag handler stores e_{id}, the element's unique id, in the only s-cache layer the new frame before pushing to $N_j.stack$ (line 12). If N_j is an SLR Head Node, the open-tag handler does not push a frame to $N_j.stack$ if the s-cache layer corresponding to N_j's tail node in the parent frame is empty, which indicates that there are *no* potential result ids waiting for the the SLR condition to be satisfied. This avoids subsequent close-tag processing for e thereby lightening the load of the close-tag handler. Once the new frame is pushed to $N_j.stack$, the result ids are moved to the t-cache layer for N_j in the parent frame (line 17).

Close-Tag Handler: If N_j is a *not* an SLR-head node, the close-tag handler moves result ids in the last (or the only) s-cache layer to the appropriate s-cache layer in the parent frame (Algorithm 2, lines 11–13). Otherwise, the result ids in the t-cache layer corresponding to tail of N_j in the parent frame are moved to the s-cache layer for N_j as they satisfy the SLR constraint at N_j (line 9). Note that between open-tag and close-tag events of a document node e that matches an SLR-head node, the potential result ids are temporarily kept in a t-cache layer for the head node. Thus the t-cache layer acts as a transit point of result ids between the open- and close-tag of e. The other option is to move result ids from one s-cache layer to the next by the open-tag handler while processing an SLR-head node. But this strategy can not be used in the presence

---

Algorithm 2. Close-Tag Event Handler

Data: Close tag $\langle /x \rangle$ of an element in the stream

1 **Close-Tag-Handler**(x)

2 $gDepth \leftarrow gDepth - 1$

3 **foreach** $N_j \in labelMap(x)$ **do**

4   **if** *(isEmpty(N_j.Stack) $\vee$ N_j.Stack[top(N_j)].levelNo $\neq$ gDepth)* **then**

5   Continue with the next iteration;

6       $F_c \leftarrow pop(N_j.stack); N_p \leftarrow parent(N_j)$;

7   **if** *SLR-Head node* **then**

8          $N_t \leftarrow$ SLR-tail(N_j);

9   Move ids in $N_p.stack[F_c.frameNo].t\text{-}cache[N_j]$ to

          $N_p.stack][F_c.frameNo].s\text{-}cache[N_j]$

10   **else if** *Result Node* **then**

11   Move ids in $F_c.s\text{-}cache[1]$ to $N_p.stack[F_c.frameNo].s\text{-}cache[N_j]$

12   **else**

13   Move ids in $F_c.s\text{-}cache[N_j.fanOut]$ to

          $N_p.stack[F_c.frameNo].s\text{-}cache[N_j]$

14   **if** *N_j has an frontier node N_e and N_e.Stack is not empty* **then**

15   **foreach** $k \in 1 \ldots N_j.fanout$ **do**

16   Move ids from $F_c.s\text{-}cache[k]$ to $N_e.Stack[top(N_e.Stack)].s\text{-}cache[1]$

17   **if** *N_j is the output node* **then**

18   Output ids in $N_p.stack[F_c.frameNo].s\text{-}cache[N_j]$

19       $Reclaim(F_c)$

---

of predicates, in which case a stack frame can be 'evaluated' only at close-tag. See Section 3.3. The frame in an SLR-parent node N_j can have result ids which have not yet satisfied the SLR conditions at the children of N_j. These ids are moved to the top stack frame of the end-node, if any (lines 14–16), as they satisfy conditions on OaT nodes *below* the frontier node but *yet to satisfy* conditions on nodes above.

Example 2. We use OaT T_{6a}, document D_2 and stack snap-shots in Fig. 5 to illustrate the algorithm. Here $\langle x \rangle$ is used to denote the frame for document node x. Fig. 5(c) shows stack contents at the opening tag of node f_1. Since there is a sequence of frames in the stacks of nodes r and b which, along with f_1, satisfies relationships along the path r–b–c–f in the query, $\langle f_1 \rangle$ is pushed to S_f. Since f is a result node, f_1 is cached to the only cache layer in $\langle f_1 \rangle$. Fig. 5(d) shows stack content after close-tag of c_2. Potential result f_1 is in $\langle b_2 \rangle.s\text{-}cache[c]$. At the close-tag of $\langle b_2 \rangle$, the result id is in $\langle b_2 \rangle.s\text{-}cache[c]$ which means that it has not yet satisfied the SLR condition from c to d as b_2 has no d-child appearing after c_2. Hence f_1 can not be moved to the cache of the frame in r. However, as c is the frontier node of b, f_1 can be moved to $\langle c_1 \rangle.s\text{-}cache[f]$ (Fig. 5(e)). Fig. 5(f) shows the stack contents at the close-tag of c_3. Both f_1 and f_2 are in $\langle b_1 \rangle.s\text{-}cache[c]$.

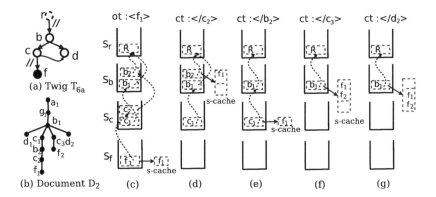

Fig. 5. Illustrating *preceding-sibling* Axis Evaluation

Subsequently, open-tag of d_2 moves the result ids from $\langle b_1 \rangle.s\text{-}cache[c]$ to $\langle b_1 \rangle.t\text{-}cache[d]$. The ids later move to $\langle b_1 \rangle.s\text{-}cache[d]$ at the close-tag of d_2 (Fig. 5(g)). The result ids finally accumulate in $\langle R \rangle.s\text{-}cache[b]$ at the close-tag of b_1.

Processing *preceding* Axis: As discussed towards the end of Section 3.1, in the OaT for an expression with *preceding* axis, the dummy root node r will have more than one branch-tree under it and the left-most tree contains the result node. The branch-trees are 'linked' by LR-edges. The algorithm for processing such an OaT has to check if the result ids obtained after processing of one branch-tree satisfy the constraints laid down by the remaining branch-trees, in left-to-right order. We can arrive at the following intuitive observation:

Observation 1. *A potential result node is said to satisfy the conditions laid down by a branch tree b with respect to the XML stream if (i) it has already satisfied the conditions of the branch trees to the left of b and (ii) a complete match for branch tree b has been found in the stream 'after' condition (i) is satisfied.*

Processing of these branch-trees are quite similar to processing of the left-most tree and is done by extending the open- and close-tag handlers to process LR-edges. Snippets 1 and 2 show these extensions. We use $R.rs\text{-}cache[N_j]$ (*resp.*, $R.rt\text{-}cache[N_j]$) to represent the rs-cache (*resp.* rt-cache) layer corresponding to the branch-tree containing a node N_j.

Close-Tag Handler: Snippet 1 shows additional steps in close-tag handler. Note that, for the left-most branch-tree, after the close-tag processing of the frame for a document node in the stack of a node N_j, the potential result ids are accumulated in appropriate s-cache layer of the parent frame (see Algorithm 2). If N_j is an LR-tail node, the accumulated ids satisfy the entire branch-tree and hence can be moved $R.rs\text{-}cache[N_j]$ (Snippet 1, line 20). For the remaining branch trees, the LR-tail processing is different. Similarly, processing of the output node needs alternative steps. We will discuss these variations slightly later.

Snippet 1. Extending Close-Tag Handler for *preceding* Axis Processing

```
----------- Replace Lines 17-19 of Algorithm 2 ----------
```
18 **if** N_j *is an LR-Tail node* **then**
19 | **if** $N_p.stack[F_c.frameNo].s\text{-}cache[N_j]$ *contains ids* **then**
20 | | Move ids in $N_p.stack[F_c.frameNo].s\text{-}cache[N_j]$ to $R.rs\text{-}cache[N_j]$
21 | **else**
22 | | **foreach** $p \in N_p.stack[F_c.frameNo].s\text{-}cache[N_j]$ **do**
23 | | | $R.rs\text{-}cache[N_j] \leftarrow R.rs\text{-}cache[N_j] \cup retrieveBucketIds(p)$

24 **if** N_j *is the output node* **then**
25 | **if** $N_p.stack[F_c.frameNo].s\text{-}cache[N_j]$ *contains ids* **then**
26 | | Output ids in $N_p.stack[F_c.frameNo].s\text{-}cache[N_j]$
27 | **else**
28 | | **foreach** $p \in N_p.stack[F_c.frameNo].s\text{-}cache[N_j]$ **do**
29 | | | $resIds \leftarrow retrieveBucketIds(p)$
30 | | | Output $resIds$

31 $Reclaim(F_c)$

Snippet 2. Extending Open-Tag Handler for *preceding* Axis Processing

```
------------ after Line 17 of Algorithm 1 ------------
```
1 **case** LR-head node
2 $N_i \leftarrow$ LR-tail(N_j);
3 **if** $R.rs\text{-}cache[N_i]$ *is not empty* **then**
4 | Move ids in $R.rs\text{-}cache[N_i]$ to a new bucket in $R.rt\text{-}cache[N_j]$
5 | Add pointer ptr to the new bucket to $F.s\text{-}cache[1]$;
6 **else if** *the last bucket in* $R.rt\text{-}cache[N_i]$ *is non-empty* **then**
7 | Add pointer ptr to the last bucket to $F.s\text{-}cache[1]$;
8 **else**
9 | Continue with the next iteration;

Open-Tag Handler: Let N_i be the tail of an LR-head node N_j. At the opening tag of an element e that matches with N_j, the potential result ids in $R.rs\text{-}cache[N_i]$ can not be moved to $R.rs\text{-}cache[N_j]$ as it is unknown whether the conditions dictated by other nodes related to N_j in the branch tree will be satisfied later by the stream. For instance, in the context of OaT T_6 and document D_1 in Fig.6, at the opening tag of h_2, it is not known if the SLR-constraint from node h to node i will be satisfied later in the stream and hence the potential result ids accumulated in $R.rs\text{-}cache[b]$ can not be moved to $R.rs\text{-}cache[h]$.

At the same time, retaining the result ids in $R.rs\text{-}cache[N_i]$ after opening tag of e is seen and moving them to $R.rs\text{-}cache[N_j]$ when the close-tag of e is seen can lead to false positives. For instance, suppose that the node id f_2 remains in $R.rs\text{-}cache[b]$ after opening-tag of h_2. When close-tag for b_3 is seen, f_4 is moved to $R.rs\text{-}cache[b]$. Subsequently, when close-tag for h_3 seen, f_4 is also moved to $R.rs\text{-}cache[h]$ with f_1. This is an error since h_3 can not decide the validity

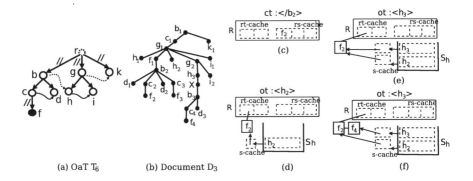

(a) OaT T_6 (b) Document D_3 (c)   (d)   (e)   (f)

Fig. 6. Illustrating *preceding* Axis Evaluation

of f_4 which is its descendant. The layered rt-cache in R can be used to avoid false positives. Intuitively, $R.rt\text{-}cache[N_j]$ acts as a temporary store for result ids in their transit from $R.rs\text{-}cache[N_i]$ to $R.rs\text{-}cache[N_j]$.

A frame has to be pushed to $N_j.stack$ in two cases – If $R.rs\text{-}cache[N_i]$ is not empty or if $R.rt\text{-}cache[N_j]$ contains at least one bucket of potential ids. The first case indicates that there are some potential results waiting for the LR-ordering constraint to be satisfied. In this case the result ids are moved to a new bucket in $R.rt\text{-}cache[N_j]$ and a pointer to this bucket is added in the s-cache of the new frame in $N_j.stack$.

A frame has to be pushed $N_j.stack$ in the second case also for the following reason: Suppose a document node x matches N_j and the $R.rs\text{-}cache[N_i]$ contains potential result ids. The ids are moved to new bucket in $R.rt\text{-}cache[N_j]$ and a pointer to this bucket is added to the s-cache of the frame for x, which is pushed to $N_j.stack$ (lines 4-5 of Snippet 1). However, at this point it is not known whether x, along with other nodes appearing in the stream will form a match for the branch tree containing N_j. At the same time another document node y appearing in the stream later may match with N_j and form, along with other nodes of the stream, a match for the branch tree of N_j *before* x can do so. Thus, a pointer to the last bucket in $R.rt\text{-}cache[N_j]$ is added in the s-cache of the frame for y. The ids in the bucket are accessed for further processing using the bucket pointer corresponding to x or y depending on which becomes part of a match for the branch tree containing N_j first (see condition (2) of Observation 1 above).

It may be observed that, an LR-head node and its branch tree are processed in the same manner as the result node and its branch-tree except that pointers to buckets in rt-cache are accumulated in s-cache of stack frames, instead of document ids. During the close-tag processing of a stack frame in an LR-tail node of these branch-trees, the accumulated bucket pointers are used to retrieve potential result ids from the rt-cache layer. The ids then are moved to rs-cache layer (line 23 of Snippet 1). The function $retrieveBucketIds(p)$ retrieves potential ids from the bucket pointed at by p and all the buckets appearing before it.

Output node processing is exactly similar except that the ids regained from buckets are output as results (line 29 of Snippet 1).

Example 3.   We use OaT T_6, document D_3, and stack snapshots in Fig. 6 to illustrate preceding axis processing. At the close-tag of b_2, f_2 moves to R.s-cache[b]. As b is an LR-tail node, f_2 further moves to R.rs-cache[b] (Fig. 6(c)). At the open-tag of h_2, the open-tag handler moves f_2 to a new bucket in R.rt-cache[h] and a pointer to the new bucket is added to $\langle h_2 \rangle$.s-cache[1] (Fig. 6(d)).

When open-tag of h_3 is seen, R.rs-cache[b] is empty, but R.rt-cache[h] contains a bucket with id f_2. Hence pointer to this bucket is added to h_3.s-cache[1] in h.stack (Fig. 6(e)). This step is essential to ensure correct computation. For instance, suppose i_1 is not present in D_3. Now, h_2 has no i-node in the stream to satisfy the SLR constraint from h to i. But h_3 has node i_2 to satisfy the said constraint and hence satisfies the branch-tree.

Eventually the pointer reaches g_2.s-cache[i] at the close-tag of i_2. At the close-tag of g_2, pointer in g_2.s-cache[i] is moved to R.s-cache[g]. Since g is an LR-tail node the pointer can be used to move contents of the bucket in R.rt-cache[h] to R.rs-cache[h]. Close-tag processing of g_1 also examines the same bucket. As the bucket is empty, no further action is needed.

(As a variation, suppose the document sub-tree containing g_2, h_3 and i_2 appears under the node labelled X after b_3. In this case the result f_4 first moves to R.rs-cache[b] and, at the open-tag of h_3, to a new bucket in R.rt-cache[h] (Fig. 6(f)). And a pointer to the new bucket appears in h_3.s-cache[1]. which eventually reaches g_2.s-cache[i] at the close-tag of i_2. At the close-tag of g_2, pointer in g_2.s-cache[i] is moved to R.s-cache[g]. Since g is an LR-tail node the pointer in can be used to move contents of both buckets in R.rt-cache[h] to R.rs-cache[h]. This is in accordance with Observation 1.)

Since k is the output node, at the close-tag of k_1, the pointer in R.s-cache[k] is used to retrieve and output the result f_2 in R.rt-cache[k].

Each rt-cache layer should be a sequence of buckets because maintaining the layer as a simple id list can lead to false positives. For instance, suppose the sub-tree rooted at X appears under g_2 to the left of h_3 in document D_3 and that i_2 is not present in the document. After close-tag processing of b_3, f_4 is cached to R.rs-cache[b]. At the open-tag of h_3, f_4 moves to R.rt-cache[h], which already contains f_2. At the close-tag of g_1, f_4 moves to R.rs-cache[h] along with f_2 which is an error. On the other hand, if R.rt-cache[h] is a linked list of buckets, f_2 and f_4 will have *separate* buckets and only f_2 will be moved to R.rs-cache[h] at the close tag of g_1.

3.3   Predicate Processing

The above algorithm can be extended to handle predicates involving *child* and *descendant* axes by modifying the stack frame structure. In OaTs representing path expressions with such predicates, the predicate expression appears as a sub-tree under the node representing the associated axis step, as in the case of conventional twigs. As predicates represent boolean conditions, they can be processed by adding a bit-vector in the stack frames of query nodes in the predicate sub-tree. Each position in the bit-vector is associated with one predicate child of the node. During query processing, a bit-vector position is asserted if the corresponding predicate child matches with a node in the stream. During close-tag event of a node, the bit-vector of the corresponding stack frame is 'evaluated'. We adopted techniques in [7] for predicate processing.

4   Experiments

In this section we compare performance of our algorithm with SPEX[2] on real world and synthetic data sets. To the best our knowledge, SPEX is the only stream query processing system that implements backward ordered axes. Java implementation of the system is publicly available (*http://spex.sourceforge.net*). Our algorithm was also implemented in Java. Xerces SAX parser from *http:// sax.sourceforge.net* was used to parse the XML documents. We ran all our experiments on a 1000 MHz AMD Athlon 3000+ machine with 2GB memory running Linux. Java virtual machine (JVM) version 1.5 was used for conducting the tests.

We used two datasets in the experiments – SWISSPROT and TREEBANK[11]. SWISSPROT is a real world dataset. TREEBANK is a deeply recursive synthetic dataset containing English sentences tagged with parts of speech.

Experiment 1: In this experiment, we compared the scalability of our system (referred as RX) with SPEX. We tested scalability with respect to increasing query size and increasing document size. Fig. 7(a) shows the test results for scalability with document size for TREEBANK dataset. Randomly generated document chunks of size (d) 0.1M, 0.2M, 0.3M and 0.4M were used and test was done for random queries of axis-step count 3 and 9. Fig, 7(b) shows test results for scalability with query size. Random queries of axis-step count (q) 3, 5, 7 and 9 were tested against documents of size 0.2M and 0.4M. Fig. 8(a) and Fig. 8(b) shows scalability results on SWISSPROT dataset ($q = 2, 4, 6, 8$). In both cases RX outperforms SPEX by wide margins. The improved performance of RX is due to its increased 'awareness' about order and effective use of that information to avoid processing of large number of elements in the stream. Note that, in RX, no frame for an SLR-head node is stacked during open-tag processing if the cache layer for the tail of the node is empty. Similar is the case with LR-head node processing. This avoids the overhead due to close-tag processing of those frames.

Experiment 2: In this experiment we examined the maximum result buffer sizes for RX and SPEX. For each algorithm, we computed the maximum of the number of result nodes maintained at various points of time during execution. This was averaged over randomly created queries. The experiments were performed on TREEBANK dataset as it is deeply recursive can lead to excessive number of potential answers. Document sizes of 0.1M, 0.3M, 0.5M and 0.7M were used in the experiment. Fig. 9 shows the result. It was found that RX uses the same amount of bufferspace as SPEX.

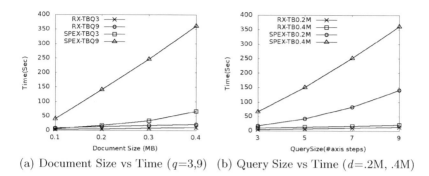

(a) Document Size vs Time (q=3,9)   (b) Query Size vs Time (d=.2M, .4M)

Fig. 7. Scalability on TREEBANK

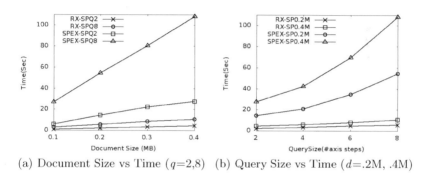

(a) Document Size vs Time (q=2,8)   (b) Query Size vs Time (d=.2M, .4M)

Fig. 8. Scalability on SWISSPROT

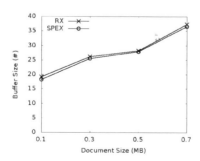

Fig. 9. Maximum Buffer Size

5   Conclusion

In this paper, we demonstrated that ordered backward axes can be effectively represented by extending conventional twigs and proposed an algorithm for processing XPath expressions with ordered backward axes against streaming data.

It was found that the algorithm is both efficient and scalable and outperforms the currently available algorithm without any additional buffer space. Explicit forward constraints for representing backward ordered axes and an effective framework for caching partial results at the query nodes have helped to detect unwanted computations during matching and lead to significant improvement in performance. It would be interesting to investigate how the current algorithm can be extended to handle bigger XPath subsets.

References

1. Barton, C., Charles, P., Goyal, D., Raghavachari, M., Fontoura, M., Josifovski, V.: Streaming XPath Processing with Forward and Backward Axes. In: ICDE, pp. 455–466 (2003)
2. Olteanu, D.: SPEX: Streamed and progressive evaluation of XPath. IEEE Trans. Knowl. Data Eng. 19(7), 934–949 (2007)
3. Abdul Nizar, M., Sreenivasa Kumar, P.: Efficient Evaluation of Forward XPath Axes over XML Streams. In: 14th International Conference on Management of Data (COMAD), pp. 217–228 (2008)
4. Raj, A., Sreenivasa Kumar, P.: Branch Sequencing Based XML Message Broker Architecture. In: ICDE, pp. 217–228 (2007)
5. Chen, S., Li, H.G., Tatemura, J., Hsiung, W.P., Agrawal, D., Candan, K.S.: Twig^{2}stack: Bottom-up processing of generalized-tree-pattern queries over XML documents. In: VLDB, pp. 283–294 (2006)
6. Chen, Y., Davidson, S.B., Zheng, Y.: An Efficient XPath Query Processor for XML Streams. In: ICDE, p. 79 (2006)
7. Gou, G., Chirkova, R.: Efficient Algorithms for Evaluating XPath over Streams. In: SIGMOD Conference, pp. 269–280 (2007)
8. Candan, K.S., Hsiung, W.P., Chen, S., Tatemura, J., Agrawal, D.: AFilter: Adaptable XML Filtering with Prefix-Caching and Suffix-Clustering. In: VLDB Conference, pp. 559–570 (2006)
9. Josifovski, V., Fontoura, M., Barta, A.: Querying XML streams. VLDB Journal 14(2), 197–210 (2005)
10. Peng, F., Chawathe, S.S.: XSQ: A streaming XPath engine. ACM Trans. Database Systems 30(2), 577–623 (2005)
11. http://www.cs.washington.edu/research/xmldatasets/

BPI-TWIG: XML Twig Query Evaluation

Neamat El-Tazi[1] and H.V. Jagadish[2]

[1] Faculty of Computers and Information, Information Systems Department,
Cairo University, Egypt
neamatab@umich.edu
[2] Electrical Engineering and Computer Science Department, University of Michigan,
Ann Arbor, MI, USA
jag@umich.edu

Abstract. We propose a new algorithm, *BPI-TWIG*, to evaluate XML twig queries. The algorithm uses a set of novel twig indices to reduce the number of comparisons needed for the twig evaluation and transform the join operation to an intersection operation between the contributing twig paths inside the query. In this paper, we present our technique and experimentally evaluate its performance.

Keywords: XML Query processing, Twig Queries.

1 Introduction

A twig pattern is a small tree whose nodes are tags, attributes or text values and edges are either Parent-Child (P-C) edges or Ancestor-Descendant (A-D) edges. Finding all the occurrences of a twig pattern specified by a selection predicate on multiple elements in an XML document is a core operation for efficient evaluation of XML queries. Indexing is often used to speed up database operations that would otherwise be expensive. However, the number of possible twig patterns is too large to be amenable to an index.

We note that the number of different path types, defined based purely on the tags at nodes along a path, is usually quite small, even if the number of actual paths is very large (and proportional to database size). This idea was exploited in our previous work, *BPI-CHAIN* in [5], to develop a new efficient structural join algorithm. However, this algorithm could address only chain queries.

The central idea of *BPI-CHAIN* no longer applies to twig queries: the number of different twig types can be very large – we get combinatorial explosion. So a direct application of the *BPI-CHAIN* idea is insufficient for the more general problem. In this paper, we develop new indices named *Twig indices*, and use these as a basis for a new algorithm to process twig queries called *BPI-Twig*.

2 Basic Data Structures

Definition 1 (Signature {s}). *The **signature** of a node is its tag encoded as an integer for convenience. The signature is used to identify the label not the node, hence, we store only distinct labels.*

Z. Bellahsène et al. (Eds.): XSym 2009, LNCS 5679, pp. 17–24, 2009.

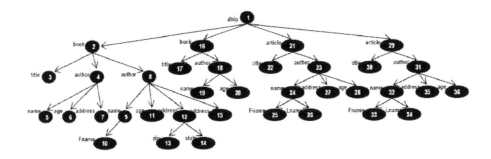

Fig. 1. XML Tree

Definition 2 (Twig {Tw}). *An XML twig, in our algorithm, is a two-level tree combining a root and its children leaves presented as $Tw(n_{root}, n_1, ..., n_m, s_{root}, s_1, ...s_m)$, where n_{root} is the twig root node number (node Identifier given to each node by traversing the XML document in depth first order), each n_i is a node number, m is the number of children in this twig and s_i is the signature of node n_i.*

a)

ID	Twig Types (explanation)	Signatures
TWT1	dblp (book, article)	s1 s2 s11
TWT2	book (title, author)	s2 s3 s4
TWT3	author(name, address, age)	s4 s5 s6
TWT4	name (fname)	s5 s8
TWT5	address(zip, state)	s7 s9 s10
TWT6	author (name, age)	s4 s5 s6
TWT7	article (title, author)	s11 s3 s4
TWT8	name (Fname, Lname)	s5 s8 s12

b)

Sig	Twig Type Ids
s1	TWT1
s2	TWT2
s4	TWT3, TWT6
s5	TWT4, TWT8
s7	TWT5
s11	TWT7

c)

Sig	Twig Type Identifiers
s3	TWT2, TWT7
s6	TWT3, TWT6
s8	TWT4, TWT8
s9	TWT5
s10	TWT5
s12	TWT8

d)

Subtree	PathTwig Type	Path Twig ID
ST1	PTWT1	Tw1 Tw2 Tw3
ST1	PTWT2	Tw1 Tw2 Tw4 Tw5
ST1	PTWT3	Tw1 Tw2 Tw4 Tw6
ST2	PTWT4	Tw1 Tw7 Tw8
ST3	PTWT5	Tw1 Tw9 Tw10
ST4	PTWT5	Tw1 Tw12 Tw13

e)

Signature	Label
s1	dblp
s2	book
s3	title
s4	author
s5	name
s6	age
7	address
8	Fname
s9	zip
s10	state
s11	article
s12	Lname

Fig. 2. a)Twig Types Index, b)Twigs Roots Index, c)Non Twig-Roots Index, d)Path Twigs Index, e) List of Signatures

Definition 3 (Twig Type {TWT}). *An XML **Twig Type** is the set of signatures $S(s_{root}, s_1, ..., s_m)$, where s_{root} is the signature of the Tw root, s_1 is the signature of the first child node of the root and s_m is the signature of the last child node of the Tw root. If duplicate signatures occur in Tw, only one of these signatures is stored in the twig type.*

In the **Twigs** index, we store distinct TWTs. There might be n number of Tw having the same TWT. The TWT guarantees that a specific Tw, having that type, contains a specific signature. We also maintain two indices **Twig Roots** and **Twig Leaf** indices to help in finding all TWTs having a query labels as their root or as their leaf in some cases. Figure 2b and c show the **Twigs-Roots** index and **Twig Leaf** index for the XML tree in Figure 1 respectively.

Definition 4 (Path Twig {PTw}). *A **Path Twig** is a set of Twigs on the same path from the XML root until the leaf nodes of the last twig on this XML path, considering only one link between any two consecutive levels in the XML tree.*

Definition 5 (Path Twig Type {PTWT}). *A **Path Twig Type** is the set of (TWT) along a (PTw).*

Definition 6 (Subtree {ST}). *A **Subtree** is a tree having one of the second level nodes of the main tree as its root. The tree in Figure 1 has four subtrees.*

In the **Path Twigs** index, we store each PTw as Tw node numbers that occur along the PTw, the key to this index is the PTWT and the ST that contains that PTw.

a)
ID	Twig Types
PTW1	TWT1 TWT2 TWT3
PTW2	TWT1 TWT2 TWT3 TWT4
PTW3	TWT1 TWT2 TWT3 TWT5
PTW4	TWT1 TWT2 TWT6
PTW5	TWT1 TWT7 TWT3 TWT8

c)
Path Twig Type	Subtrees (array)
PTW1	ST1
PTW2	ST1
PTW3	ST1
PTW4	ST2
PTW5	ST3, ST4

b)
ID	Signatures	Nodes
Tw1	s1 s2 s2 s11 s11	n1 n2 n16 n21 n29
Tw2	s2 s3 s4 s4	n2 n3 n4 n8
Tw3	s4 s5 s6 s7	n4 n5 n6 n7
Tw4	s4 s5 s6 s7 s7	n8 n9 n11 n12 n15
Tw5	s5 s8	n9 n10
Tw6	s7 s9 s10	n12 n13 n14
Tw7	s2 s3 s4	n16 n17 n18
Tw8	s4 s5 s6	n18 n19 n20
Tw9	s11 s3 s4	n21 n22 n23
Tw10	s4 s5 s7 s6	n23 n24 n27 n28
Tw11	s5 s8 s12	n24 n25 n26
Tw12	s11 s3 s4	n29 n30 n31
Tw13	s4 s5 s7 s6	n31 n32 n35 n36
Tw14	s5 s8 s12	n32 n33 n34

Fig. 3. a)Path Twig Types Index, b)Twig IDs c)Path twig-Types - Subtrees Index

Path Twig Types Index. In this index, each PTWT is stored as an array of TWT identifiers that occur along that path.

Path Twig Types - Subtrees Index. In order to reduce intermediate results, we store the PTWT identifiers with the ST identifier so that, if we have multiple PTWTs inside a query, we can intersect the STs of the PTWT to deduce which subtrees that contain all these query PTWT. This will reduce the number of subtrees fetched to perform the join.

3   Twig Query Evaluation Algorithm

To evaluate twig queries, we introduce the *BPI-Twig* algorithm. The input to the algorithm is a query tree. The algorithm finds all Tws that match the input query tree and outputs all node numbers contained inside those matched Tws. Figure 4 presents all functions of the algorithm with their inputs and outputs.

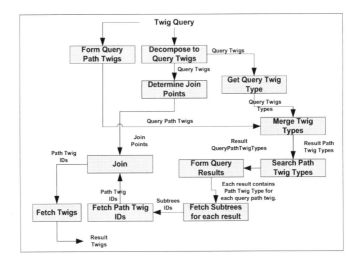

Fig. 4. BPI-TWIG Evaluation Functions

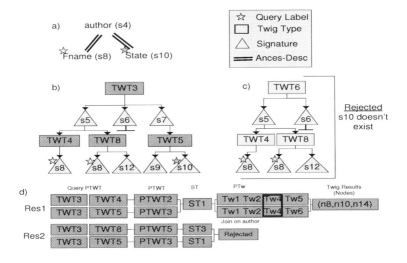

Fig. 5. Twig Query Evaluation Example

Example 1. Consider the twig query pattern author[.//Fname][.//state] presented as a query tree in Figure 5a. The signatures of the three tags author, Fname and state are s4, s8 and s10 respectively. Since the query Tw is only one level twig, there is no decomposition involved. There is one join at the root(author) level. Searching for the author(s4) in the **Twig Roots** index returns two query TWTs which are TWT3 and TWT6. In Figure 5b, we use TWT3 as a starting point in our search and in Figure 5c, we use twig type *TWT6* as a starting point. We search for s8 and s910 inside the signatures of each *TWT*.

In Figure 5b, the two signatures s8 and s10 do not exist in TWT3. Therefore, we search for TWTs having the signatures of TWT3 as their roots in the **Twig Roots** index. The result is TWT4 and TWT8 for s5, TWT5 for s7.

For each output TWT, we still searching for signatures s8 and s10, we find s8 in TWT4 and TWT8. And signature s10 is found in TWT5. Since the two signatures s8 and s10 are found as descendants to TWT3, therefore all the shaded TWTs under TWT3 are considered in producing the result. On the other hand, in Figure 5c, the same search is done but only s8 is found, therefore we do not consider *TWT6* a part of the result. We then merge the shaded TWTs into query PTWT according to the results in each *TWT*. The first result of this query pattern will contain the two query PTWT (TWT3, TWT4) and (TWT3, TWT5). Query PTWT (TWT3, TWT4) contains s8(Fname) and (TWT3, TWT5) contains s10(state). The second result contains the two query PTws(TWT3, TWT8) and (TWT3, TWT5).

For each query result returned, we search for the PTWTs that have the same TWT of the query PTWT in the **Path Twig Types Index**. For the first result {(TWT3, TWT4), (TWT3, TWT5)}, the PTWT that contain each of the query PTWT are PTWT2, that contains the twig types (TWT1, TWT2, TWT3, TWT4), and PTWT3 that contains the twig types (TWT1, TWT2, TWT3, TWT5). Both PTWTs have to be in the same ST. This is because the query root is not the same as the XML document root. Therefore by searching for STs for PTWT2 and PTWT3 in **Path Twig Types -Subtrees** index, we find that both PTWTs occur in the same subtree, ST1. On the other hand, the second result {(TWT3,TWT8), (TWT3,TWT5)} is contained in PTWT5 and PTWT3. These PTWTs occur in ST3 and ST1 respectively. The result is rejected since there is no ST intersection.

For the passing result that contains PTWT2 and PTWT3, we search for PTw Identifiers that has these types in the intersection ST1, using **Path Twigs** index, the resulting PTws are Tw1, Tw2, Tw4, Tw5 and Tw1, Tw2, Tw4, Tw6 as shown in the second and third row of Figure 2d. These two PTws has to be joined on the same position of the query root TWT. The third position is the join position which can be known from the position of the query signature within the TWT. Since the Tw in the third position in both PTws is the same, Tw4, therefore, these two PTws join successfully and can be added to the output Tw identifiers.

To output the result we start from the Tw identifier that has the same signature as the query root, Tw4. We fetch Tw4 from **Twig Structure**, and output its root node n8, then fetch the next Tw in the first PTw {Tw1, Tw2, Tw4, Tw5} which is Tw5, and output the node number that has the same signature **Fname**, s8, node n10. Then fetch the last Tw from the second PTw {Tw1, Tw2, Tw4, Tw6}, Tw6, and output the node number that has the same signature **state**, s10, node n14. Therefore the output of the query is the triplet (n8, n10, n14).

An interesting point here is that if a query Tw has greater depth, it does not mean that the number of joins increases. For instance, consider the following query: //dblp/book[./title] [./author[./name] [./address[./city]

[./state][./zip]] [./publisher[./name] [./phone[./areacode]
[./number]]]]. This query has only two query PTws and it contains only
a **single join** on the node(book).

4   Performance Evaluation

In this section, we present results of experiments to caompare the performance
of the proposed algorithm, *BPI-TWIG* with the state of the art *LCS-TRIM*
algorithm [3]. We chose *LCS-TRIM* as one base line because it is regarded as
the "best" structural join algorithm proposed thus far. We are grateful to the
inventors of *LCS-TRIM* for sharing their code with us to enable this comparison.
Both algorithms were implemented in C++ and all experiments were run on a
2GHz Intel CPU with 1 GB RAM running WinXP.

We used a popular XML data set: DBLP [7]. The data set characteristics are
presented in Table 1. As we can see, the size of the bitmap index structure is only
a small fraction of the size of the data set. In other words, the space overhead
for the *Path Twig Types - Subtrees* index to support *BPI-TWIG* is negligible.
Number of path twig types is very small compared to the number of twigs in
the XML document. The *Twig IDs* index stores the node numbers for each twig.
We do not access this index except to get the result nodes from the result path
twigs. Indeed, most of the evaluation time is taken in reporting node numbers to
the output. Against the data set we ran a number of queries shown in Table 2.

Table 1. Data set characteristics

DataSet size	TWIG IDs Index	Subtree Index	Other Indices Total
DBLP 170MB	70MB	22MB	4MB
#Subtrees	#TwigTypes	#PathTwigTypes	#Path Twigs
331285	328858	752	331285

Table 2. DBLP Twig Query Set

Q#	Query Expression
Q1	//phdthesis[./year][./number]
Q2	//phdthesis[./year][./series][./number]
Q3	//book[./author][./cdrom]
Q4	//inproceedings[./author][./month]
Q5	//book[./author][./publisher]

Figure 6 presents the evaluation times for queries over the *DBLP* data set
using both *BPI-TWIG* and *LCS-TRIM*. We find that the improvement of *BPI-
TWIG* over *LCS-TRIM* for most queries is more than 99%. The output results
in these queries ranges from 1 to 1200 results. The improvement in Q5 is 98%
which is less than the other improvements because Q5 has approximately 1200
which takes more time than other queries in retrieving the twig node numbers.
It is clear that *BPI-TWIG* out performs *LCS-TRIM* in the figure.

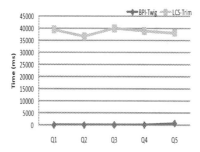

Fig. 6. DBLP Twig Queries Evaluation

4.1 Query Complexity Effect and Recursiveness Effect

Figure 7 is obtained by running against a synthetic data set generated using *Toxgene* [8]. Q5 in the figure, has the form $B[E][F[G][H[I[P][Q[R][S]]]]]$. The first four queries were prefixes of this query. The output size is constant. We have seen in [5] how the length of chain makes a little difference to query evaluation time. In Figure 7, time to execute queries increases linearly with number of branching points (twigs) in the query.

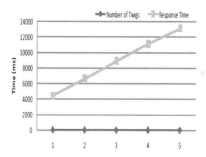

Fig. 7. Change in BPI-TWIG evaluation time with Query Complexity

Since most of the computation is performed on the types level not the data level, the effect of recursiveness on the query processing is not significant. On the other hand, the existence of recursiveness might help more our approach by minimizing the number of twig types available in the indexes. In addition to minimizing the response time due to the smaller number of twig types available in the indexes.

5   Related Work and Conclusion

Recently, new approaches have been introduced to evaluate twig queries such as *VIST* [2], *PRIX* [4] and *LCS-TRIM* [3]. These approaches rely on subsequence matching for the query sequence inside the data sequence to get the result of the join. In addition to the high maintenance cost for the built indices, other drawbacks comes from the data representation itself. For instance, the worst case storage requirement in *VIST* for a B-tree index used is higher than linear in the total number of nodes of the XML document.

In this paper, we proposed a new technique, *BPI-TWIG*, to evaluate twig queries in XML. *BPI-TWIG* maintains a small bit-mapped data structure with which make it able to do very fast in-memory processing of parent-child and ancestor-descendant structural joins within twig queries. The bulk of the time in *BPI-TWIG* is spent in simply reading out the results. In other words, structural join computation is essentially as fast as index look up, but without the need to maintain unwieldy join indices. Our preliminary evaluation shows the algorithm to be a viable solution to the twig query evaluation problems. The ideas presented in this paper are very simple, but very effective. We hope that they can transform how real systems compute structural joins in twig queries.

References

1. Jagadish, H.V., Al-Khalifa, S., Chapman, A., Lakshmanan, L.V.S., Nierman, A., Paparizos, S., Patel, J.M., Srivastava, D., Wiwatwattana, N., Wu, Y., Yu, C.: TIMBER: A Native System for Querying XML. VLDB Journal 11(4), 274–291 (2002)
2. Wang, H., Park, S., Fan, W., Yu, P.S.: Vist: A dynamic index method for querying xml data by tree structures. In: SIGMOD Conference, pp. 110–121 (2003)
3. Tatikonda, S., Parthasarathy, S., Goyder, M.: LCS-Trim: Dynamic programming meets xml indexing and querying. VLDB, 63–74 (2007)
4. Rao, P., Moon, B.: Prix: Indexing and querying xml using prüfer sequences. In: ICDE, pp. 288–300 (2004)
5. El-Tazi, N., Jagadish, H.V.: BPI: XML Query Evaluation using Bitmapped Path Indices. In: DATAX (2009)
6. El-Tazi, N.A., Jagadish, H.V.: Xml query evaluation using bitmapped path and twig indices, http://www.eecs.umich.edu/db/BPI/appendix.pdf
7. DBLP computer science bibliography dataset, http://kdl.cs.umass.edu/data/dblp/dblp-info.html
8. Barbosa, D., Mendelzon, A.O., Keenleyside, J., Lyons, K.A.: ToXgene: An extensible template-based data generator for XML. In: WebDB, pp. 49–54 (2002)

On the Efficiency of a Prefix Path Holistic Algorithm*

Radim Bača and Michal Krátký

Department of Computer Science, Technical University of Ostrava
Czech Republic
{radim.baca,michal.kratky}@vsb.cz

Abstract. In recent years, many approaches to XML twig pattern searching have been developed. Holistic approaches such as TwigStack are particularly significant in that they provide a powerful theoretical model for optimal processing of some query types. Holistic algorithms use various partitionings of an XML document called *streaming schemes* and they prove algorithm optimality depending on query characteristics.

In this article, we introduce a variant of the TwigStack algorithm which can work with various streaming schemes. Its efficiency does not deteriorate when the number of streams per query node is increased, as it does in the case of the iTwigJoin algorithm. Since the indices utilized by the iTwigJoin and our algorithm are exactly the same, we can use heuristics to select the appropriate algorithm. The aim of this paper is to show that the prefix path streaming scheme algorithms can be efficient even for documents with many labeled paths.

1 Introduction

Recently, many approaches to the indexing of the XML data structure have been developed [6, 18, 4, 1]. Twig pattern queries (TPQ) represent an important part of the XPath and XQuery [17] languages used for XML data querying. We can find some works integrating TPQ into XQuery algebra [12]. This is important since we can not expect that we will process each query using a single algorithm.

There are many works comparing various algorithms and indices for TPQ searching [18, 9, 8, 1, 4, 11]. These works show that the inverted list with a special purpose algorithm can speed up the query processing for many queries. We can find different types of inverted lists depending on the key utilized. The most common key is the element's name [18, 9, 8, 1, 4] or the element's root-to-node labeled path [15, 13, 11, 5, 3]. During the TPQ processing where labeled paths are used, we have to first match the TPQ in a DataGuide tree [2]. We retrieve the set of the inverted list's keys (labeled paths in this case) for each query node. A TPQ algorithm using labeled paths can become quite inefficient as the size of these sets grows [5].

---

* Work is partially supported by Grants of GACR No. 201/09/0990 and IGA, FEECS, Technical University of Ostrava, No. BI 4569951, Czech Republic.

Z. Bellahsène et al. (Eds.): XSym 2009, LNCS 5679, pp. 25–32, 2009.

In [14], we can find a comparison of various approaches to TPQ processing based on structural joins [18, 1], holistic joins [4, 5], and on sequence searching. Holistic approaches were considered the most robust solution requiring no complicated query optimization. Moreover, holistic approaches provide a powerful theoretical background for optimal processing of some query types. This is quite useful during the query processing since we can precisely determine the TPQ processing complexity.

In this work, we address a problem of the iTwigJoin+PPS holistic algorithm using labeled paths [5]. We show that our new algorithm is very useful when we have a higher number of keys per query node during the query processing. Since the indices utilized by both algorithms are the same, we can employ heuristics to select the appropriate algorithm. We show that the prefix path streaming algorithms can be efficient even for documents with many labeled paths.

This paper is organized as follows: In Section 2, we depict a model of an XML document. Section 3 briefly describes the previously published theory behind holistic approaches. In Section 4, we introduce our new holistic algorithm called TwigStackSorting. Section 5 shows that our new algorithm is efficient even for documents with many labeled paths.

2   Model

It is common to model an XML document as an *XML tree*, where the tree nodes correspond to elements and attributes of an XML document. In what follows, we shall simply write 'node' instead of the correct 'tree node'. We can see an example of the XML tree in Figure 1(a). Nodes in this XML tree are pre-order numbered for easy reference in the following examples.

For each node n of an XML tree we define a *labeled path* as a sequence $tag_0/tag_1/\ldots/tag_n$ of node tags lying on a path from the root to n. Every labeled path occurs only once in a DataGuide tree [15].

Join algorithms are usually based on a labeling scheme, where a label is assigned to every node of an XML tree. Node label enables us to determine the relationship between two nodes. Containment labeling scheme [18] is a frequently used labeling scheme, which is also utilized in holistic approaches considered in this article [4, 5].

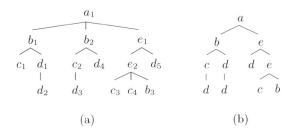

(a)   (b)

Fig. 1. (a) XML tree (b) DataGuide of the XML tree

2.1   Twig Query Pattern and Query Matching

A TPQ can be modeled as an unordered rooted query tree, where each node of the query tree corresponds to a single query node, and the edges represent an AD or PC relationship between the connected nodes. Query match of a TPQ in an XML tree is one occurrence of the pattern in the XML tree. The solution is one root-to-leaf path in a query match.

Example 1. (TPQ Examples) Let us have the TPQ Q //e[./c]//b. The query matches of Q in the XML tree in Figure 1(a) are (e_2, c_3, b_3), (e_2, c_4, b_3). Solutions of the path //e/c are (e_2, c_3) and (e_2, c_4).

TPQ can be understood as a single operator which is a part of an XQuery algebra used by a query processor. In [12], a comprehensive attempt to incorporate the TPQ into the XQuery algebra has been made. This algebra uses TPQs having a single output query node. Similarly, XPath query language specifies a single output query node in a TPQ. We can observe that even in the case of FLWOR XQuery expression, where more than one query node of the twig is in the output, the twig usually contains only one iterative query node (bounded with the for clause). Therefore, the output of a TPQ processing is not a set of query matches (as it is proposed in [4,5]), but the algorithm output includes a set of node labels corresponding to the output query node. In many cases, this slight difference has a significant impact on the efficiency of a holistic join algorithm. Let us discuss this issue in more detail in the next section.

3   Holistic Algorithms

In this section, we briefly introduce holistic algorithms for a TPQ searching [4, 5, 10]. Holistic approaches use an abstract data type (ADT) called a *stream*. A stream is an ordered set of equivalent node labels. The most common equivalences (also known as *streaming schemes* [5]) are defined according to the node name (tag streaming) or node labeled path (prefix path streaming - PPS).

A cursor pointing to the first node label is assigned to each stream. The following operations are defined for the stream T:

- *head(T)* – returns the node label corresponding to the cursor's position
- *eof(T)* – returns true if the cursor is at the end of T
- *advance(T)* – moves the cursor to the next node's label

Let us note that the stream ADT is often implemented by an inverted list data structure.

In any algorithm using the prefix path streaming, we must first find labeled paths matching the query in the DataGuide tree. Since the DataGuide is a tree as well, we can use some tag streaming holistic methods for the searching [2]. By PRU_q we denote a set of labeled paths matching the query node q in the DataGuide.

Example 2. Let us consider the XML document in Figure 1(a) and the TPQ //b[./c]//d. Under LPS the $PRU_b = \{T_{a/b}\}$, $PRU_c = \{T_{a/b/c}\}$, and $PRU_d = \{T_{a/b/b/d}, T_{a/b/d}\}$.

3.1 Holistic Algorithms

Holistic algorithms work in two phases:

- **First phase** – the algorithm scans whole streams and prunes useless nodes. Nodes matching the TPQ are stored temporarily in stacks and solutions are stored in persistent arrays during this phase.
- **Merge phase** – the algorithm merges solutions (stored in the persistent arrays), prunes irrelevant solutions and outputs query matches.

In every step of the first phase, the algorithm checks all the query node's streams and searches for the nodes which match the query. The TwigStack algorithm [4] has only one stream per query node, therefore, it decides whether the node matches the query or is useless by using the *head(T)* node of each TPQ's query node stream. In the case of iTwigJoin [5], we have a set of streams per query node and thus the situation is more complicated. The iTwigJoin algorithm uses a *soln(T_{lp}, q')* set, which is a set of streams T_{lp} of class q, where the relationship between lp and lp' satisfies the relationship between q and q'.

Sometimes during one step, the algorithm is not able to determine a stream, where the head node is a part of a query match or is useless. We say that the algorithm is optimal when the situation never occurs during the first phase. The TwigStack and iTwigJoin+PPS algorithms are optimal for a TPQ having only AD edges and iTwigJoin+PPS is optimal also for a query with one branching node.

In connection with the issues mentioned in Section 2, we do not return query matches. Therefore, the second phase becomes obsolete in the case of an optimal holistic algorithm. An optimal holistic algorithm finds only relevant nodes and returns nodes corresponding to the output query node found in the first phase, therefore, the merge phase can be skipped.

Holistic approaches described in [4,5,10] also perform the second phase in the case of an optimal algorithm because they return the result in a form of query matches. In order to process the second phase efficiently we have to sort the solutions founded in the first phase. Holistic approaches describe the *node blocking method* to sort the solutions without excessive sorting algorithms. However, the node blocking can lead to a repeated read and write of the same pages in the secondary storage and it can be particularly inefficient when many nodes are blocked.

4   The TwigStackSorting Algorithm

In this section, we describe the TwigStackSorting algorithm. The TwigStackSorting modifies the TwigStack [4] in order to use the PPS scheme. This means that TwigStackSorting is an alternative approach to iTwigJoin+PPS [5] since they both use the same index structure.

The basic TwigStack algorithm remains the same with the only difference being in the *advance(q)* and *head(q)* methods. Compared to the TwigStack algorithm, we have more than one stream per query node. In this case, we have to keep streams for each query node q sorted according to the streams' head. As usual, we implement this by using an array of references to the streams' head. The $advance(q)$ method first shifts the head of the current stream to the next node. Therefore, the array of references must be re-sorted since it must handle the order of streams sorted according to the streams' head. The $head(q)$ function of the query node q simply selects the head of the lowest stream in the array.

Basically we merge streams corresponding to one query node into one stream. However, this merging is simply performed during the TwigStackSorting run.

4.1   Analysis of the TwigStackSorting Algorithm

The correctness of the TwigStackSorting algorithm can be shown analogously to TwigStack due to the fact that they both use the same stack mechanism. Therefore, the space, time, and I/O complexity is the same as for TwigStack if the algorithm is optimal. Due to this fact, TwigStackSorting is optimal for queries having only AD axes. Let us note that I/O is usually lower for a PPS algorithm than for TwigStack since the TPQ match in a DataGuide filters out many irrelevant streams.

The iTwigJoin+PPS and TwigStackSorting algorithms use the same input streams, however their performance can be significantly different. iTwigJoin+PPS often searches the minimal value in the *soln* set and this operation has to be performed with a time consuming sequence scan within the *soln* set of streams. On the other hand, TwigStackSorting uses the binary search algorithm during the $advance(q)$ method with logarithmic complexity.

4.2   Prefix Path Streaming Optimality

Let us define a *checking query node* q of a TPQ as a query node having an AD relationship with its parent query node and having a PC relationship in a subtree. For example, the TPQ //a/b[/c]//d[//e]//f has single checking node a and the TPQ //a/b[//c/d]/d/f has two checking nodes a and c.

It can be shown that a PPS algorithm is optimal when PRU_q set of every checking query node q does not contain two labeled paths, where one is prefix of another one. This optimality condition may be applied to iTwigJoin+PPS as well as to TwigStackSorting. Proof of this is out of the scope of this paper.

Example 3. (PPS optimality) Let us consider the TPQ //b[/c]/d and the XML tree in Figure 1(a). In the case of TwigStack, we find the first output node d_1 which is a part of the query match (b_1, c_1, d_1). During the following steps, node d_2 is skipped and streams' cursors are moved to nodes b_2, c_2, and d_3. The nodes do not form a query match, however we are not able to decide whether any node matches the TPQ or is useless without a stream scan. In this situation, TwigStack is blocked and it has to output the useless solution (b_2, d_3), which will be pruned during the second phase of the algorithm.

Such a situation does not occur in the case of TwigStackSorting. We first search the labeled paths which match the TPQ in the DataGuide from

Figure 1(b). These labeled paths are a/b, $a/b/c$, $a/b/d$. One for each query node. Therefore, the PRU_d set includes only the stream $T_{a/b/d} = \{d_1, d_4\}$ and the problematic d_3 node is omitted.

5   Experimental Results

In our experiments[1], we compare TwigStackSorting with iTwigJoin+PPS. We implemented these approaches in C++. We process experiments with the INEX 1.9 collection [7] which contains a higher number of labeled paths, therefore, issues of iTwigJoin+PPS discussed in Section 4.1 can be clearly shown here.

5.1   Prefix Path Streaming Optimality

Table 1 shows the statistics of the various XML collections [16, 7]. The third column shows the number of the *PPS optimal tags*. The PPS optimal tag x is a tag which never has a tag x as an ancestor (i.e., it is not recursive). If the TPQ has non-recursive nodes as checking query nodes, then the PPS holistic algorithm will be optimal for such a TPQ. We can see that the number of optimal tags is significant in each XML document. Therefore, the optimality mentioned in Section 4.2 covers the significant number of queries.

Table 1. Statistics of various collections

Collection	LP count	Tag count	PPS optimal tags	
			Count	Ratio
TreeBank	338,749	251	234	93.33%
XMARK	548	548	75	97.40%
INEX-wiki	114,870	3,608	3,435	95.21%
INEX-1.9	16,018	217	190	87.56%
DBLP	170	41	39	97.56%
Nasa	110	69	69	100%
SwissProt	264	99	99	100%

5.2   Processing Time Results

We randomly selected 25 twig queries having six query nodes with a combination of PC and AD edges. We chose PPS optimal queries with various number of labeled paths per query node. Each query was processed three times with a cold cache and then the average processing time per kilobyte read from the secondary storage was computed.

In Figure 2, we show the dependency of the processing time on the number of labeled paths. We also depict the fitting curve. In Figure 2(a), we see the problem of iTwigJoin+PPS mentioned in [5]. Its processing time per kilobyte grows

---

[1] The experiments were executed on Intel Pentium 4 1.66Ghz, 2.0 MB L2 cache; 2GB 667MHz DDR2 SDRAM; Windows XP.

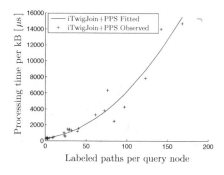
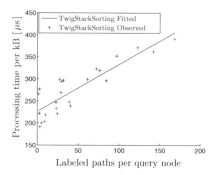

Fig. 2. Processing time depending on the number of labeled paths (a) iTwigJoin+PPS (b) TwigStackSorting

quadratically with the number of labeled paths per query nodes. TwigStackSorting is significantly more robust as we see in Figure 2(b).

There can be queries which are optimal for iTwigJoin+PPS, but they are not optimal for TwigStackSorting. However, we experimentally evaluated that the iTwigJoin algorithm is usually faster only for queries having less then five labeled paths per query node.

6   Conclusion

We show that the holistic algorithm using labeled paths is very efficient even for a higher number of labeled paths per query node. TwigStackSorting is shown to be more robust and its performance does not deteriorate with an increase in the number of labeled paths. We also discuss the optimality of PPS algorithms since the PPS algorithms can be optimal for a significant number of twig queries. This can be important for a query optimizer, which can precisely determine the query processing cost. In our future work, we want to formalize the proposed PPS optimality.

References

1. Al-Khalifa, S., Jagadish, H.V., Koudas, N.: Structural Joins: A Primitive for Efficient XML Query Pattern Matching. In: Proceedings of ICDE 2002. IEEE CS, Los Alamitos (2002)
2. Bača, R., Krátký, M.: On the Efficient Search of an XML Twig Query in Large DataGuide Trees. In: Proceedings of the Twelfth International Database Engineering & Applications Symposium, IDEAS 2008. ACM Press, New York (2008)
3. Bača, R., Krátký, M.: TJDewey – On the Efficient Path Labeling Scheme Holistic Approach. In: Proceedings of Database Systems for Advanced Applications, DASFAA 2009 International Workshops. Springer, Heidelberg (to appear, 2009)
4. Bruno, N., Srivastava, D., Koudas, N.: Holistic Twig Joins: Optimal XML Pattern Matching. In: Proceedings of ACM SIGMOD 2002, pp. 310–321. ACM Press, New York (2002)

5. Chen, T., Lu, J., Ling, T.: On Boosting Holism in XML Twig Pattern Matching Using Structural Indexing Techniques. In: Proceedings of ACM SIGMOD 2005, pp. 455–466. ACM Press, New York (2005)
6. Florescu, D., Kossmann, D.: Storing and Querying XML Data using an RDMBS. IEEE Data Engineering Bulletin 22(3), 27–34 (1999)
7. Fuhr, N., Gövert, N., Malik, S., Lalmas, M., Kazai, G.: INEX (2007), http://inex.is.informatik.uni-duisburg.de/2007/
8. Grust, T., van Keulen, M., Teubner, J.: Staircase Join: Teach a Relational DBMS to Watch Its (Axis) Steps. In: Proceedings of VLDB 2003, pp. 524–535 (2003)
9. Tatarinov, I., et al.: Storing and Querying Ordered XML Using a Relational Database System. In: Proceedings of ACM SIGMOD 2002, New York, USA, pp. 204–215 (2002)
10. Lu, J., Ling, T., Chan, C., Chen, T.: From Region Encoding to Extended Dewey: on Efficient Processing of XML Twig Pattern Matching. In: Proceedings of VLDB 2005, pp. 193–204 (2005)
11. Yoshikawa, T.S.M., Amagasa, T., Uemura, S.: XRel: a Path-based Approach to Storage and Retrieval of XML Documents Using Relational Databases. ACM Trans. Inter. Tech. 1(1), 110–141 (2001)
12. Michiels, P., Mihaila, G., Simeon, J.: Put a tree pattern in your algebra. In: Proceedings of the 23th International Conference on Data Engineering, ICDE 2007, pp. 246–255 (2007)
13. Milo, T., Suciu, D.: Index structures for path expressions. In: Beeri, C., Bruneman, P. (eds.) ICDT 1999. LNCS, vol. 1540, pp. 277–295. Springer, Heidelberg (1999)
14. Moro, M., Vagena, Z., Tsotras, V.: Tree-pattern Queries on a Lightweight XML Processor. In: Proceedings of VLDB 2005, pp. 205–216 (2005)
15. Goldman, J.W.R.: DataGuides: Enabling Query Formulation and Optimization in Semistructured Databases. In: Proceedings of VLDB 1997, pp. 436–445 (1997)
16. University of Washington's Database Group. The XML Data Repository (2002), http://www.cs.washington.edu/research/xmldatasets/
17. W3 Consortium. XQuery 1.0: An XML Query Language, W3C Working Draft (November 12, 2003), http://www.w3.org/TR/xquery/
18. Zhang, C., Naughton, J., DeWitt, D., Luo, Q., Lohman, G.: On Supporting Containment Queries in Relational Database Management Systems. In: Proceedings of ACM SIGMOD 2001, pp. 425–436 (2001)

KSRQuerying: XML Keyword with Recursive Querying

Kamal Taha and Ramez Elmasri

Department of Computer Science and Engineering,
The University of Texas at Arlington, USA
{kamal.taha,elmasri}@uta.edu

Abstract. We propose an XML search engine called KSRQuerying. The search engine employs recursive querying techniques, which allows a query to query the results of a previous application of itself or of another query. It answers recursive queries, keyword-based queries, and loosely structured queries. KSRQuerying uses a sort-merge algorithm, which selects subsets from the set of nodes containing keywords, where each subset contains the *smallest* number of nodes that: (1) are *closely* related to each other, and (2) contain at least one occurrence of each keyword. We experimentally evaluated the quality and efficiency of KSRQuerying and compared it with 3 systems: XSeek, Schema-Free XQuery, and XKSearch.

Keywords: XML, keyword search, loosely structured search, recursive querying.

1 Introduction

With the emergence of the World Wide Web, online businesses, and the concept of ubiquitous computing, business' XML databases are increasingly being queried directly by customers. Business' customers and employees may not be fully aware of the exact structure of the underlying data, which prevents them from issuing structured queries. Keyword-based querying does not require the user to be aware of the structure of the underlying data nor elements' label. Bur, the precision of results could be low. On the other hand, Loosely Structured querying allows combining some structural constraints within a keyword query, by specifying the context where a search term should appear (combining keywords and element names). That is, it requires the user to know *only* the labels of elements containing the keywords, but does not require him to be aware of the structure of the underlying data. Thus, Loosely Structured querying combines the convenience of Keyword-Based querying while enriching queries by adding structural conditions, which leads to performance enhancement. We propose in this paper an XML search engine called KSRQuerying. The search engine employs recursive querying techniques, which allows a query to query the results of a previous application of itself or of another query. The search engine answers recursive queries, keyword-based queries, and loosely structured queries.

Extensive research has been done in XML keyword querying [13, 14, 21]. Computing the Lowest Common Ancestor (LCA) of elements containing keywords is the common denominator among most proposed search engines. Despite the success of

Z. Bellahsène et al. (Eds.): XSym 2009, LNCS 5679, pp. 33–52, 2009.

these search engines, they suffer *recall* and *precision* limitations. The reason is that these engines employ mechanisms for building relationships between data elements based solely on their labels and proximity to one another while overlooking the *contexts* of the elements. The context of a data element is determined by its parent, because a data element is usually a characteristic of its parent. If for example a data element is labeled title, we cannot determine whether it refers to a book title or a job title without referring to its parent. KSRQuerying employs context-driven search techniques to avoid the pitfalls of non context-driven systems. The engine also uses a stack sort-merge algorithm, which selects subsets from the set of nodes containing keywords, where each subset contains the *smallest* number of nodes that contain at least one occurrence of each keyword. KSRQuerying enables the user to issue a query based on his *degree of knowledge* of the underlying data as follows: (1) if the user knows only keywords, he can submit a keyword-based query, (2) if he is unaware of the structure of the underlying data, but is aware of the elements' labels he can submit a loosely structured query, and (3) if he is unaware of the structure of the underlying data and the query requires *transitive closure* of a relation, he can submit a recursive query.

The rest of the paper is organized as follows. In section 2 we present related work. In section 3 we present definitions of key concepts used in the paper. In section 4, we present our context-driven search techniques. In section 5 we show how to select from the set of nodes containing keywords subsets that are *closely* related to each other. In sections 6 and 7 we show how answer nodes are determined. In section 8, we present the system implementation and architecture. We present the experimental results in section 9 and our conclusions in section 10.

2   Related Work

Researchers in the area of keyword search in relational databases [2, 3, 10] consider the relational database as a graph, where nodes represent the tuples in the database and edges represent the relationships between the nodes. The result of a query is a sub-graph that contains all the query's keywords. A number of studies [4, 5, 8] propose modeling XML documents as graphs, and keyword queries are answered by processing the graphs based on given schemas.

We proposed previously two XML search engines called OOXSearch [17] and CXLEngine [18]. They differ from KSRQuerying in that: (1) they answer only loosely structured queries, (2) they do not answer recursive queries, (3) they employ Object Oriented techniques while KSRQuerying employ stack based sort-merge algorithm, and (4) they may return results with low recall/precision for queries that expand across an XML tree. The studies [13, 14, 21] are related to this paper, since they propose semantic search techniques for establishing relationships between nodes in XML documents modeled as trees. Despite their success, however, they suffer recall and precision limitations as a result of overlooking nodes' contexts. We take [13, 14, 21] as samples of non-context driven search engines and overview below the techniques employed by each of them. *We compared the three systems experimentally with KSRQuerying (see section 10).*

XSeek [14]: [14] uses the approach of XKSearch for identifying search predicates by determining SLCA (which [14] calls it *VLCA nodes*). The contribution of [14] is the inference of result nodes that are *desirable* to be returned. Each desirable node is a data node that either: (1) matches one of the query's return nodes (if the label of node n_1 matches keyword k_1 and there does not exist a descendant node n_2 of n_1 whose label matches another keyword k_2, n_1 is considered a return node), or (2) matches a predicate (a keyword that does not satisfy the condition in (1) is a predicate).

XKSearch [21]: [21] returns a subtree rooted at a node called the Smallest Lowest Common Ancestor (SLCA), where the nodes of the subtree contain all the query's keywords and they have no descendant node(s) that also contain all the keywords. Consider for example Fig. 2 and let node 8 contains the keyword "XQuery" instead of "XML and the Web". Now consider the query Q("XQuery", "Wilson"). Even though the keyword "XQuery" is contained in both nodes 8 and 15, the answer subtree will be the one rooted at node 10 (which is the SLCA of nodes 11 and 15) and not the one rooted at node 7 (which is the LCA of nodes 11 and 8).

Schema-Free XQuery [13]: In [13], nodes **a** and **b** are *not meaningfully related* if their Lowest Common Ancestor (LCA), node **c** is an ancestor of some node **d**, which is a LCA of node **b** and another node that has the same label as **a**. Consider for example nodes 2, 11, and 15 in Fig. 2. Nodes 15 title and 2 name are not related, because their LCA (node 1) is an ancestor of node 10, which is the LCA of nodes 11 and 15, and node 11 has the same label as node 2. Therefore, node 15 is related to node 11 and not to node 2 (name).

3 Concepts Used in KSRQuerying

In this section we present definitions of key notations and basic concepts used in the paper. We model XML documents as rooted and labeled trees. A tree t is a tuple $t = (n, e, r, \lambda t)$ where n is the set of nodes, $e \subseteq n \times n$ is the set of edges, r is the root node of t, and $\lambda t : n \to \Sigma$ is a node labeling function where Σ is an alphabet of node labels. A node in a tree represents an element in an XML document. Nodes are numbered for easy reference. We use the abbreviation IAN throughout the paper to denote "Intended Answer Node". An IAN is a requested return node containing the data that the user is looking for, where the data is relevant to the query's keywords.

Definition 3.1. *Ontology Label (OL) and Ontology Label Abbreviation (OLA):* Let $m \to m'$ denote that class m is a subclass of class m' in an Object-Oriented Ontology. This relationship could be expressed also as m "is-a" m' e.g. a customer "is a" person. An *Ontology Label* m' of a node m, where m is an interior node label in the XML tree, is the most general superclass (root node) of m in a defined ontology hierarchy. Fig.1 shows an example of ontology hierarchy. The Ontology Label of an interior node m is expressed as $OL(m) = m'$. Since customer→person in Fig.1, the Ontology Label of node customer(1) in Fig. 2 is expressed as OL(customer) = person. Taxonomically, m' is a cluster set that contains entities sharing the same domain, properties, and cognitive characteristics e.g. cluster person contains the entities customer, author, etc. For each entity (interior node in an XML tree) $m \in m'$, $OL(m) = m'$. The framework of KSRQuerying applies the above mentioned clustering

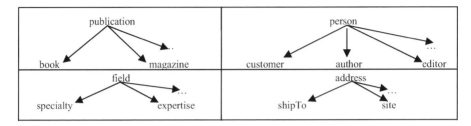

Fig. 1. Example of ontology hierarchy

Table 1. OLs and OLAs of the parent nodes in Fig. 2

Parent nodes (with their IDs)	OL
customer(1), author (10), editor(34), processer(22)	person
book(7), magazine(31), latestPublication(14)	publication
expertise(12), field(17), specialty (36)	field
shipTo (19), site (24)	address
currentOrder (3), previousOrder (27)	order

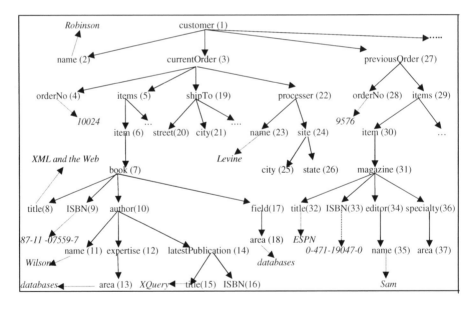

Fig. 2. Customer publication order (file: order.xml)

concept to all parent nodes in an XML tree, and the label of each of these clusters is an Ontology Label (OL). Table 1 shows the Ontology Labels and clusters of parent nodes in the XML tree in Fig. 2. The table is an alternative representation of the information in Fig.1.

Definition 3.2. *Canonical Tree (CT):* Let (n', n) denote that there is an edge from node n' to node n in the XML tree. We call n the child of n'. A *Canonical Tree* T is a pair, $T = (OL(n'), N)$, where $OL(n')$ is the Ontology Label of an interior node n' and N is a finite set of leaf data nodes and/or attributes: $N = \{ n \mid n$ is a leaf data node and (n', n), or n is an attribute of $n' \}$. In Fig. 2 for example, the parent node customer(1) and its leaf child data node name(2) constitute a Canonical Tree, and node customer(1) is represented in the Canonical Tree by its Ontology Label person (see the root Canonical Tree T_1 in Fig. 3). A Canonical Tree is the simplest semantically meaningful subtree. A data node by itself is an entity that is semantically meaningless. The Ontology Label of a Canonical Tree is the Ontology Label of the parent node component of the Canonical Tree. For example, the Ontology Label of Canonical Tree T_1 in Fig. 3 is the Ontology Label of the parent node customer (1), which is person. A Canonical Tree is represented by a rectangle. The label above the rectangle, which has the form T_i, represents the numeric ID of the Canonical Tree, where $1 \le i \le |T|$. For example, in the *Canonical Trees Graph* (see Definition 3.3) shown in Fig. 3, the Ontology Label of the root Canonical Tree is person and its numeric ID is T_1. We use the abbreviation "CT" throughout the paper to denote "Canonical Tree".

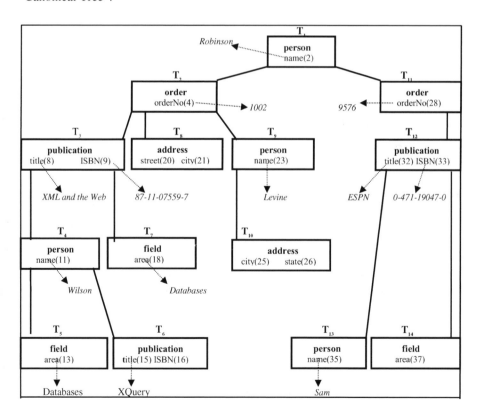

Fig. 3. Canonical Trees Graph of the XML tree in Fig. 2

Definition 3.3. *Canonical Trees Graph (CTG):* A Canonical Trees Graph *CTG* is a pair of sets, *CTG= (T_S , E)*, where T_S is a finite set of CTs and *E*, the set of edges, is a binary relation on T_S, so that $E \subseteq T_S \times T_S$. $T_S = \{T_i \mid T_i \text{ is a CT, and } 1 \le i \le \mid T_S \mid\}$. The CTG is constructed as follows: If the two interior nodes *n*, *n'* in the XML tree both have leaf data nodes and/or attributes, and either (1) (*n*, *n'*) is an edge in the XML tree, or (2) *n* is an ancestor of *n'* in the XML tree, and there does not exist any node *n''* on the path from *n* to *n'* where *n''* has leaf data nodes or attributes, then (T_1,T_2) will be an edge in the CTG, where $T_1 = (OL(n),N_1)$ and $T_2 = (OL(n'),N_2)$ and N_1 is the set of leaf children data nodes/attributes of *n* and N_2 is the set of leaf children data nodes/attributes of *n'* . In Fig. 2 for example, since node book(7) is a descendant of node currentOrder(3) and there is no interior node in the path from book(7) to currentOrder(3) that has a child data node(s)/attribute(s), CT T_3 is a child of CT T_2 (see Fig. 3).

Definition 3.4. *Keyword Context (KC):* it is a CT containing one or more of a query's keywords.

Notation 3.1. OL_T: OL_T denotes the Ontology Label of CT *T*. In Fig. 3 for example OL_{T_1} is person.

4   Determining the Immediate Relatives of Canonical Trees

Notation 4.1 Intended Answer Node (IAN)
When a user submits a query, he is usually looking for data that is relevant to the query's keywords. We call each one of the data nodes containing this data an Intended Answer Node (IAN). Consider for example Fig. 2 and that the user submitted the keyword-based query *Q* ("XQuery"). As the semantics of the query implies, the user wants to know information about the book, whose title is "XQuery" (node 15). This information is contained in nodes 11 and 16. Thus, each of nodes 11 and 16 is considered an IAN.

We call each CT *T* that *can* contain an IAN for a KC an Immediate Relative (IR) of the KC. We denote the set of CTs that are Immediate Relatives of the KC by IR$_{KC}$. Consider for example Figures 2 and 3 and the keyword-based query *Q*("XQuery"). XQuery is a publication's title contained in node 15. It is intuitive that data node 16 and/or 11 be an IAN, but it is not intuitive that data node 2 be an IAN, because "Robinson" did not order this publication. Since "XQuery" is contained in T_6, we can determine that each of the CTs containing nodes 16 and $11 \in IR_{T_6}$ while the CT containing node $2 \notin IR_{T_6}$.

Table 2. Abbreviations of concepts used in the paper: *Abr.* denotes abbreviation

Abr.	Concept	Abr.	Concept	Abr.	Concept
IAN	Intended Answer Node	LCA	Lowest Common Ancestor	CTG	Canonical Trees Graph
CT	Canonical Tree	OL	Ontology Label	KC	Keyword Context
IR$_T$	Immediate Relatives of CT *T*	RKC	Related Keyword Contexts	OLA	Ontology Label Abbreviation

Proposition 4.1. For CT T to be an Immediate Relative of a KC, its Ontology Label should be different than both OL_{KC} and $OL_{T'}$, where T' is a CT located between T and the KC in the CTG.

Proof (heuristics)
a) Since each IR of a KC can contain an IAN for the query, we are going to prove heuristically that if $OL_T \neq OL_{KC}$, then CT T can contain an IAN; otherwise, it cannot. Let T_i and T_j be two distinct CTs having the same Ontology Label. Therefore, the two CTs share common entity characteristics and some of their data nodes are likely to have the same labels. Let n_1, n_2, n_3, n_4, n_5, and n_6 be data nodes, where n_1, n_2, $n_3 \in T_i$ and n_4, n_5, $n_6 \in T_j$. Let n_1 and n_4 have the same label l_1, n_2 and n_5 have the same label l_2, n_3 has the label l_3, and n_6 has the label l_4. Let $d_{m'}^m$ denote the distance between data nodes m and m' in the XML tree. Now consider the query $Q(l_1 = "k_i"$, $l_2?)$, and that the keyword "k_i" is contained in data node $n_1 \in T_i$ (the KC is T_i) and l_2 is the label of the IAN. Intuitively, the IAN is $n_2 \in T_i$ and not $n_5 \in T_j$, because $d_{n_2}^{n_1} < d_{n_5}^{n_1}$.

If the label of the IAN in the same query is l_3 (instead of l_2), then obviously the IAN is $n_3 \in T_i$. However, if the label of the IAN in the same query is l_4, then the query is meaningless and unintuitive. Now consider the query $Q(l_3 = "k_i"$, $l_1?)$. Intuitively, the IAN is n_1 and not n_4 due to the proximity factor. If the label of the IAN in the same query is l_2 (instead of l_1), intuitively the IAN is n_2 and not n_5. Thus, we can conclude that in order for the query to be meaningful and intuitive, the IAN cannot be contained in CT T_j if the keyword is contained in CT T_i. In other words, an IAN of a query cannot be contained in a CT whose Ontology Label is the same as that of the KC.

b) Let: (1) CT $T' \in IR_{KC}$, (2) T' be a descendant of the KC, and (3) CT T be a descendant of T'. In order for T to be an IR of the KC, intuitively T has to be an Immediate Relative of T', because T' relates (connects) T with the KC. If T and T' have the same Ontology Label, then $T \notin IR_{T'}$ (according to the conclusion of heuristics a); therefore, $T \notin IR_{KC}$. Thus, in order for CT T' to be an Immediate Relative of the KC, $OL_T \neq OL_{T'}$. We now formalize the concept of Immediate Relatives in Definition 4.1.

Definition 4.1. *Immediate Relatives of a KC (IR_{KC})*
The Immediate Relatives of a KC is a set IR_{KC}, $IR_{KC} = \{ T' \mid T'$ is a CT whose Ontology Label is different than OL_{KC} and $OL_{T'}$, where T' is a CT located between T and the KC in the CTG$\}$.

Proposition 4.2. *If CT $T \notin IR_{KC}$ and CT T' is related (connected) to the KC through T, then CT $T' \notin IR_{KC}$.*

Proof (induction): Every CT T has a domain of influence. This domain covers CTs, whose *degree of relativity* to T is strong. Actually, these CTs are the Immediate Relatives of T. If CT $T' \notin IR_T$, then the degree of relativity between T' and T is weak. Intuitively, the degree of relativity between any other CT T'' and T is even weaker if T'' is connected to T through T', due to proximity factor.

We can determine IR_{KC} by pruning from the CTG all CTs $\notin IR_{KC}$, and the remaining ones would be IR_{KC}. We present below three properties that regulate the pruning process. Properties 1 and 2 are based on Proposition 4.1 and property 3 is based on Proposition 4.2.

Property 1: When computing IR_{KC}, we prune from the CTG any CT, whose Ontology Label is the same as the Ontology Label of the KC.

Property 2: When computing IR_{KC}, we prune CT T' from the CTG if: (1) there is another CT T'' between T' and the KC, and (2) the Ontology Label of T'' is the same as that of T'.

Property 3: When computing IR_{KC}, we prune from the CTG any CT that is related (connected) to the KC through a CT T, $T \notin IR_{KC}$.

Example 1: Let us determine $IR_{T_{12}}$ (recall Fig. 3). By applying property 2, CT T_2 is pruned because it is located in the path T_{12}, T_{11}, T_1, T_2 and its Ontology Label is the same as the Ontology Label of CT T_{11}, which is closer to CT T_{12}. By applying property 3, all CTs that are connected with CT T_{12} through CT T_2 are pruned. The remaining CTs in the CTG are $IR_{T_{12}}$ (see Fig. 4-A).

Example 2: Let us determine IR_{T_6}. By applying property 1, CT T_3 is pruned because its Ontology Label is the same as that of CT T_6. By applying property 3, all CTs that are connected with CT T_6 through CT T_3 are pruned. The remaining CTs in the CTG are IR_{T_6} (see Fig. 4-B).

Example 3: Figs. 4-C, D, E, and F show IR_{T_1}, IR_{T_9}, IR_{T_7}, and IR_{T_3} respectively.

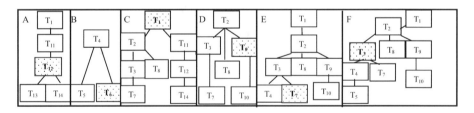

Fig. 4. A) $IR_{T_{12}}$ B) IR_{T_6} C) IR_{T_1} D) IR_{T_9} E) IR_{T_7} F) IR_{T_3}

We constructed an efficient algorithm called ComputeIR (see Fig. 5) for computing IR_{KC}. To compute IR_{KC}, instead of examining each CT in the graph, we only examine the CTs that are *adjacent* to any CT $T' \in IR_{KC}$. That is, if the algorithm determines that some CT $T' \in IR_{KC}$, it will then examine the CTs that are adjacent to T'. However, if the algorithm determines that $T' \notin IR_{KC}$, it will not examine any CT T'' that is connected to the KC through T', because T'' is known to be not an IR of the KC (according to property 3). The algorithm's time complexity is $O(\sum_{i=1}^{|T|} |IR_{T_i}|)$.

```
ComputeIR (KC) {
1. T ← KC
2. S_{KC}^{KC} ← null
3. IR_{KC} ← null
4. ExamineCT (T) {
5.     for each CT T' ∈ Adj [T] {
6.         if ( OL_{T'} ≠ OL_{KC}  &  OL_{T'} ∉ S_{KC}^{T} )
7.             then { IR_{KC} = IR_{KC} ∪ T'
8.                     S_{KC}^{T'} ← S_{KC}^{T} ∪ OL_{T'}
9.                     ExamineCT (T')
                } /*end if */
        } /*end for */
} /* end  ExamineCT */
} /* end the algorithm */
```

Fig. 5. Algorithm ComputeIR

5   Determining Related KCs

We select from the KCs subsets, where each subset contains the *smallest* number of KCs that: (1) are *closely related* to each other, and (2) contain at least one occurrence of each keyword. The KCs contained in each of these subsets are called *Related Keyword Contexts (RKC)*.

Definition 5.1. *Canonical Relationship Tree*
The Canonical Relationship Tree of CTs T and T' (denoted by $R_{T,T'}$) is the set of CTs in the CTG located in the path from CT T to CT T' including CTs T and T'. For example, the Canonical Relationship Tree of CTs T_1 and T_{14} in Fig. 3 is the set $\{T_1, T_{11}, T_{12}, T_{17}\}$.

Let s be the set of CTs containing the search terms of a query (the KCs). Let the subset $\{ T, T' \} \subseteq S$, and that this subset contains at least *one occurrence of each keyword*. The KCs contained in set $R_{T,T'}$ (the Canonical Relationship Tree of CTs T and T') *collectively* constitute the Related Keyword Contexts (RKC), if either: (1) the CTs in set $R_{T,T'}$ have *distinct* Ontology Labels, or (2) only T and T' have the *same* Ontology label, which is *different* than the Ontology Labels of the other CTs in $R_{T,T'}$. If set S contains n subsets satisfying the conditions specified above, there will be n RKCs. We now formalize the concept of RKC in definition 5.2.

Definition 5.2. *Related Keyword Contexts (RKC)*
RKC is a set of CTs, where: (1) for each two distinct CTs $T_i, T_j \in$ RKC, $T_i \in IR_{T_j}$, (2) the CTs in set RKC contain at least one occurrence of each keyword, and (3) there are no two CTs $T_i, T_j \in$ RKC, where T_i and T_j contain the same keywords.

We now present Example 4 to illustrate the RKC concept.

Example 4: Consider Figures 2 and 3 and the keyword-based query: Q ("Levine", "databases", title). The keyword "databases" is contained in nodes 13 and 18, and the label title matches nodes 8, 15, and 32. We show below how by employing the RKC concept we would be able to select from the set of nodes containing the keywords and the label, the subset 8, 18, and 23 as the smallest number of nodes that are *closely* related to one another and contain the two keywords and the label. As can be seen, the semantics of the answer is correct, because the employee "Levine" (node 23) processed the order of the publication whose title is contained in node 8 and the field area of this publication is "databases" (node 18). On contrast, the Stack Algorithm of [21] would answer the same query incorrectly by returning the set of nodes 15, 18, and 23 and considering node 3 as their *SLCA*. As can be seen the semantics of the answer is incorrect, because "Levine" (node 23) did not process an order of a publication whose title is contained in node 15. The reason for the faulty answer is that the Stack Algorithm of [21] does not employ context-driven search techniques. Keyword "databases" is contained in CTs T_5 and T_7. The label 'title' is contained in CTs T_3, T_6, and T_{12}. Keyword "Levine" is contained in CT T_9. Each of the following Canonical Relationship Trees contains at least one occurrence of each keyword:

> R_{T_9,T_3,T_5} = {T_9, T_2, T_3, T_4, T_5}. The relationship tree contains CTs T_4 and T_9, which have the same Ontology Labels. Therefore, the KCs in the set do not constitute RKC.

> R_{T_9,T_3,T_7} = {T_9, T_2, T_3, T_7}. There are no two or more CTs in the set having the same Ontology Label. Therefore, the KCs T_9, T_3, and T_7 constitute RKC.

> R_{T_9,T_6,T_5} = {T_9, T_2, T_3, T_4, T_5, T_6}. The relationship tree contains CTs T_4 and T_9, which have the same Ontology Labels. Therefore, the KCs in the set do not constitute RKC.

> R_{T_9,T_6,T_7} = {T_9, T_2, T_3, T_7, T_4, T_6}. The relationship tree contains CTs T_4 and T_9, which have the same Ontology Labels. Therefore, the KCs in the set do not constitute RKC.

> R_{T_9,T_{12},T_7} = {T_{12}, T_{11}, T_1, T_2, T_9, T_3, T_4, T_5}. The relationship tree contains more than two CTs having the same Ontology Label. Therefore, the KCs in the set do not constitute RKC.

> R_{T_9,T_{12},T_5} = {T_{12}, T_{11}, T_1, T_2, T_9, T_3, T_7}. The relationship tree contains more than two CTs having the same Ontology Label. Therefore, the KCs in the set do not constitute RKC.

Thus, from the set of nodes containing the keywords and the label, the subset 8, 18, and 23 are the smallest number of nodes that are *closely* related to one another and contain the two keywords and the label.

6   Locating an IAN for a Recursive Query

KSRQuerying answers XML queries that require recursion using recursive querying technique, which allows a query to query the results of a previous application of itself or of another query. A query can be composed of more than one component. Each component is composed of FROM, WHERE, and RETURN clauses. Fig. 6 shows a form of a recursive query. The WHERE clause can contain one or more search predicates connected using the n-ary connectives *and*, *or*, and *not*. A query can contain variable names, each preceded by the construct '*var*'. A variable acts as *handle* for the data it is bound to. In the query form shown in Fig. 6, variable X is bound to a node labeled *label$_1$*. This node contains data, which would be passed to query component Q_2. The node is contained in a CT T, $T \in \bigcap_{T' \in RKC} IR_{T'}$, where T' is one of the query's KCs belonging to set RKC. That is, the node is contained in a CT located in the intersection of the IRs of the CTs composing the query's RKCs. Each IR is computed using the three pruning properties described in section 4. The expression KC(*var X*) in query component Q_2, denotes: the CT containing the node bound by variable X is the KC. Variable Y is bound to the data contained in the IAN. This IAN is located in a CT $\in IR_{KC}$.

Q_1:
FROM   "*XML doc*"
  WHERE k_1 and k_2 and...k_n and var X = *label$_1$*
RETURN   *TempResultConstruct* (*var X*)

Q_2:
FROM   *TempResultConstruct* (*var X*), "*XML doc*"
WHERE   *KC (var X)*   and   var Y = *label$_2$*
RETURN   *FinalResultConstruct* (*var X*) [*var Y*]

Fig. 6. Form of a recursive query

Example 5: Consider Fig. 2 and that while "Levine" (node 23) was processing an order of a publication (whose subject area is "databases"), he discovered that he had been making an error in all the previous orders, whose subject areas is "databases". Therefore, he decided to identify the names of the customers who placed these orders to notify them. So, he constructed the recursive query shown in Fig. 7. Variable X in query component Q_1 is bound to the titles of the publications: (1) whose subject areas is "databases", and (2) which are processed by "Levine". Using the technique described in section 5 for computing RKC, we will find that RKC ={T_9, T_7}. Thus, variable X will be bound to the value of node n labeled "title", where $n \in \{ IR_{T_7} \cap IR_{T_9} \}$. Recall Figures 4-E and 4-D for $IR_{T_7} \cap IR_{T_9}$. Variable X will be bound to the value of a node labeled "title" and contained in a CT located in: $\{ IR_{T_7} \cap IR_{T_9} \} = \{ T_2, T_3, T_8, T_{10} \}$. Thus, variable X will be bound to node 8 contained in T_3. Q_2 queries the results returned by Q_1 to determine the names of the customers, who ordered the publications, whose titles are the value of node 8. The KC in Q_2 is CT T_3. Variable Y is bound to the IAN name, which will be located in a CT $\in IR_{T_3}$. Recall Fig. 4-F for IR_{T_3}. The IAN is node 2 contained in CT T_1.

Q_1:
FROM   "file: order.xml"
WHERE "Levine" *and* "databases" *and* var X = title
RETURN   publicationsTitles (*var X*)

Q_2:
FROM   publicationsTitles (*var X*), "file: order.xml"
WHERE   KC (*var X*) *and*   var Y = name
RETURN   NameOfCustomers [*var Y*]

Fig. 7. Recursive query of Example 5

Example 6: Consider Fig. 2 and that the publication distributor wants to suggest books for customers after their current orders are completed. The distributor believes that a customer is likely to be interested in books that were previously ordered by the author of the book, which the customer is currently ordering. That is, books that were ordered by the author (node 10), who ordered them in a role of a customer (node 1). The distributor submitted the recursive query shown in Fig. 8. Q_1 will return the name of the author of the book, which the customer is currently ordering. Q_2 will return the titles of the books that were previously ordered by this author. The KC of Q_1 is CT T_3 and variable X will be bound to node 11 contained in $T_4 \in IR_{T_3}$ (IR_{T_3} = {T_1, T_2, T_4, T_5, T_7, T_8, T_9, T_{10}}). Variable X will be bound to the data of node 11 in each tuple containing the title of the book, which the customer is currently ordering (node 8). In Q_2, the CT containing node "customer/name", (which is T_1) is the KC. Variable Y is bound to the IAN "node 8" contained in $T_3 \in IR_{T_1}$.

Q_1:
FROM   "file: order.xml"
WHERE   book/title   *and*   var X = author/name
RETURN   AuthorInfo (*var X*)

Q_2: FROM AuthorInfo (*var X*), "file: order.xml"
WHERE KC (customer/name = *var X*) *and*   var Y = book/title
RETURN   RecommendedBooks [*var y*]

Fig. 8. Recursive query of example 6

7   Constructing the Answers for Loosely Structured and Keyword Queriess

7.1   Forming an Answer Subtree for a Keyword-Based Query

The answer subtree for a keyword-based query is composed from the following CTs: (1) the RKC, and (2) each CT T_i, $T_i \in \bigcap_{T_j \in RKC} IR_{T_j}$ (the intersection of the Immediate

Relatives of the KCs composing the RKC). That is, the answer subtree is formed from the RKC in addition to each CT T_i, where T_i is an Immediate Relative of *each KC* $\in$ RKC. This methodology of constructing an answer subtree guarantees that *each* node in the answer subtree is semantically related to *all* nodes containing the keywords and to *all* other nodes in the subtree. Thus, the described methodology avoids returning results with low precision, which is one of the pitfalls of non context-driven systems. There could be more than one answer subtree for a query. If there are n RKCs, there would be n answer subtrees.

Example 7: Consider again the keyword-based query: Q ("Levine", "databases", title), which we presented in Example 4. The RKC = $\{T_9, T_7, T_3\}$. The answer subtree is composed of the RKC and the result of the intersect operation: $IR_{T_9} \cap IR_{T_7} \cap IR_{T_3}$. Recall Figs. 4-D, 4-E, and 4-F for IR_{T_9}, IR_{T_7}, and IR_{T_3} respectively. $IR_{T_9} \cap IR_{T_7} \cap IR_{T_3} = \{T_2, T_8, T_{10}\}$. So, the answer subtree is composed of the set $\{T_9, T_7, T_3, T_2, T_8, T_{10}\}$. Let us call this set S. The answer subtree, is formed from the nodes components in each CT $\in S$ in addition to connective interior nodes.

7.2 Locating an IAN of a Loosely Structured Query

The key difference between keyword-based queries and loosely structured queries lies in their search terms. The search term of the former is a keyword "k", and *each* node containing k is considered when computing RKC. The search term of the later is a label-keyword *pair* (l = "k"), and *only* nodes whose labels is l and containing the keyword k are considered when computing RKC. Thus, loosely structured querying restricts the search. Consider for example Fig. 2 and consider that node 8 contains the title "databases" instead of "XML and the Web". If a keyword-based query contains the keyword "databases", then nodes 8, 13, and 18 will be considered when computing RKC. If a loosely structured query contains the search term (title = "databases"), only node 8 will be considered when computing RKC, since the label of nodes 13 and 18 is not title. KSRQuerying answers a loosely structured query as follows. If there is only *one* node matches *each* search term, the RKC will be composed of the CTs containing these nodes (if there are n search terms, the RKC will be composed of n CTs). Otherwise, KSRQuerying will use the approach described in section 5 for determining RKC. After determining RKC, an IAN will be contained in a CT T_i, $T_i \in \bigcap_{T_j \in RKC} IR_{T_j}$ (T_i is an Immediate Relative of *each* CT $\in$ RKC).

Example 8: Consider Figures 2 and 3 and the loosely structured query: Q (ISBN ="87-11-07559-7", ISBN = "0-471-19047-0", name?). The query asks for the name of the customer, who ordered publications, whose ISBNs are 87-11-07559-7 and 0-471-19047-0. Only node 9 matches the search term (ISBN = "87-11-07559-7") and only node 33 matches the search term (ISBN = "0-471-19047-0"). Thus, the RKC will be composed of CTs T_3 and T_{12}, which contain nodes 9 and 33 respectively. The IAN name should be located in the intersect $IR_{T_3} \cap IR_{T_{12}} = T_1$ (recall Fig. 6-F for IR_{T_3} and Fig. 6-A for $IR_{T_{12}}$). The IAN is node 2 contained in T_1.

8   System Implementation and Architecture

Fig. 9 shows KSRQuerying system architecture. The XML schema describing the structure of the XML document is input to the **OntologyBuilder**, which outputs to the **GraphBuilder** the list of Ontology Labels corresponding to the interior nodes in the XML schema. The **OntologyBuilder** uses an ontology editor tool to create ontologies and populate them with instances. We used Protégé ontology editor [16] in KSRQuerying prototype system. Using the input XML schema and the list of Ontology Labels, the **GraphBuilder** creates a CTG, using Algorithm **BuildCTreesGraph** (see Fig. 10). Using the input XML document, the CTG, and the query's set of keywords the **KCdeterminer** locates the KCs. The **IRdeterminer** uses algorithm **ComputeIR** (recall Fig. 5) to compute for each CT T in the CTG its IR_T and saves this information in a hash table called IR_TBL for future references. When **KSRQuerying Query Engine** receives a query, it computes its RKC, and it then accesses table IR_TBL to construct the answer. The query engine extracts the data contained in each answer data node $n \in IR_{KC}$ using XQuery Engine [23].

8.1   Determining Ontology Labels

There are many ontology editor tools available that can be used for determining the Ontology Labels of nodes. [15] lists these tools. We used Protégé ontology editor [16] in the prototype implementation of KSRQuerying. It allows a system administrator to build taxonomies of concepts and relations and to add constraints onto domains of relations. We experimented with KSRQuerying using a large number of XML docs from INEX [11, 12] and XMark [22], and module **OntologyBuilder** (recall Fig. 9) created an Ontology Label for each *distinct* tag name. We used Protégé for creating about 25% of the ontologies and the other 75% were available in electronic form (done by others) and we imported them to the system using namespaces co-ordination. For each Ontology Label OL_i, KSRQuerying stored in a table called **OL_TBL** all tag names whose Ontology Label is OL_i (e.g. table 1).

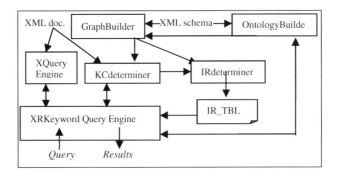

Fig. 9. XRKeyword system architecture

8.2 Constructing Canonical Trees Graphs

Algorithm BuildCTreesGraph (see Fig. 10) constructs CTGs for XML trees. Its input is the OL_TBL table (recall section 8.1) and the list of nodes adjacent to each node in the XML tree. For example, the adjacency list of node 3 in Fig. 2 is nodes 1, 4, 5, 19, and 22. Lines 1-12 construct the individual CTs and lines13-18 connect them by edges. Line 5 iterates over the nodes that are adjacent to an interior node n. If n' is a leaf data node (line 6), this node will be contained in set T_Z, which represents CT T_Z (line 8 or 12). Function setParentComp in line 9 sets node n as the parent node component of CT T_Z. Line 10 stores all the parent nodes components of CTs in set ParentComps. Function getCT in line 15 is input the closest ancestor interior node m' to interior node m, and it then outputs the numeric ID of the CT, whose parent node component is m'. The same function in line 16 outputs the numeric ID of the CT, whose parent node component is m. Function setCTparent in line 17, connects the two CTs that were output in lines 15 and 16 by an edge (setting parent-child relationship). Line 18 sets the OL of m as the OL of CT T_y.

```
BuildCTreesGraph {
1. z = 0
2. for each node  n  ∈ OL_TBL {
3.     flag = 0
4.     z = z + 1
5.     for each node  n′ ∈ adj [ n ]   {
6.         if (isLeafNode ( n′ ) = true) {
7.             then if (flag = 0)  {
8.                 then {  T_Z = T_Z ∪  n′
9.                     setParentComp ( n ,T_Z)
10.                    ParentComps = ParentComps + n
11.                    flag = 1   }
12.                else  T_Z = T_Z ∪ n′ }/*end if*/ }/*end if*/
            }/*end for*/
        }/*end for*/
13. for each node m ∈ ParentComps {
14.     m′ ← Closest ancestor node to m in set ParentComps
15.     T_x ← getCT ( m′ )
16.     T_y ← getCT (m)
17.     setCTparent (T_y, T_x)
18.     setOL (getOL (m), T_y)
    } /*end for */
  } /*end the algorithm*/
```

Fig. 10. Algorithm BuildCTreesGraph

9 Experimental Results

We have implemented KSRQuerying in Java and ran on an AMD Athlon XP 1800+ processor, with a CPU of 1.53 GHz and 736 MB of RAM, under Windows XP. We experimentally evaluated the quality and efficiency of KSRQuerying and compared it with XSeek [14], Schema-Free XQuery [13], and XKSearch [21]. The implementation of Schema-Free XQuery [13] has been released as part of the TIMBER project [20]. We used TIMBER for the evaluation of [13]. We implemented the system of XKSearch [21] from scratch. As for XSeek [14], since it uses the same approach of XKSearch [21] for identifying search predicates, we implemented it by expanding the implementation of XKSearch to incorporate XSeek's techniques that inference *desirable* nodes.

9.1 Recall and Precision Evaluation

We evaluated the quality of results returned by KSRQuerying by measuring its recall and precision and comparing it with [13, 14, 21]. We used the test data of INEX 2005 and 2006. KSRQuerying prototype system created an Ontology Label for each *distinct* tag name in the test collections used in the experiments (recall section 8.1).

Some of the documents in the INEX 2005 [11] test collection are scientific articles (marked up with XML tags) from publications of the IEEE Computer Society covering a range of topics in the field of computer science. There are 170 tag names used in the collection. On average an article contains 1,532 XML nodes, where the average depth of an element is 6.9. *We used in the experiments a set of 60 queries, with query numbers 210-269.*The test collection of INEX [12] is made from English documents from Wikipedia project marked up with XML tags. On average an article contains 161.35 XML nodes, where the average depth of an element is 6.72. *We used in the experiments a set of 40 queries, with query numbers 340-379.*

There are two types of topics (i.e. queries) included in the INEX test collections, *Content-and-structure (CAS)* queries and *Content-only (CO)* queries. All topics contain the same three fields as traditional Information Retrieval (*IR*) topics: title, description, and narrative. The title is the actual query submitted to the retrieval system. The description and narrative describe the information need in natural language. The difference between the *CO* and *CAS* topics lies in the topic title. In the case of the *CO* topics, the title describes the information need as a small list of keywords. In the case of *CAS* topics, the title describes the information need using descendant axis (//), the Boolean *and/or*, and *about* statement (it is the *IR* counterpart of *contains* function in XPath). *CAS* topics are loosely structured queries while *CO* queries are keyword-based queries.

An INEX assessment records for a given topic and a given document, the degree of relevance of the document component to the INEX topic. A component is judged on two dimensions: *relevance* and *coverage*. Relevance judges whether the component contains information relevant to the query subject and coverage describes how much of the document component is relevant to the query subject. We compared the answers obtained by each of the 4 systems to the answers deemed relevant by an INEX assessment. For a given topic and assessment, we measured how many of the XML nodes that are deemed relevant in the assessment are *missing* (for determining recall) and how many *more* XML nodes are retrieved (for determining precision). Fig. 11 shows the *average* recall and precision of the 4 systems using the 2005 and 2006 INEX test collections. As the Figure shows, the recall and precision of KSRQuerying outperform those of [13, 14, 21], which we attribute to KSRQuerying's computation of IR_{KC} and RKC and to the fact that the other 3 systems do not employ context-driven search techniques. We reached this conclusion after observing that the recall and precision of the 3 systems drop in each test data containing more than one element having: (1) the same label but representing different types, (2) different labels but representing the same type, and/or (3) a query's search term has multiple matches. The tests results showed that XSeek and XKSearch have the same recall, which is due to the fact that XSeek uses the same approach of XKSearch for identifying search predicates (see Figs. 11-a and 11-c). However, the tests results showed that the precision of XSeek outperform XKSearch, which is due to XSeek's inference mechanism for determining *desirable* result nodes. The reason that the recall of [13] outperforms [14] and [21] is because the technique it uses for building relationships is based on the hierarchical relationships between the nodes, which alleviates node labeling conflicts.

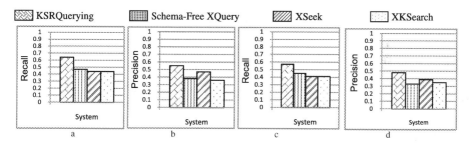

Fig. 11. (a) and (b) avg recall and precision of KSRQuerying, [13], [14], and [21] on INEX 2005. (c) and (d) avg recall and precision of KSRQuerying, [13], [14], and [21] on INEX 2006.

We show below sample of the queries used in the experiments and show how [13] returned faulty answers. We also show how KSRQuerying answered the same queries correctly. First, recall section 2 for the technique used by [13].

- Consider Fig. 12-A and the query Q (title = "Introduction", image?). The query asks for the image presented in the section titled "Introduction" (node 3). The correct answer is node 6. But, [13] returned null. The reason is that the LCA of nodes 3 and 6 is node 2, and node 2 is an ancestor of node 4, which is the LCA of nodes 6 and 5, and node 5 has the same label as node 3. Therefore, [13] considered node 6 is related to node 5 and not to node 3.

KSRQuerying answer: Let T denote a CT, whose nodes components are nodes 2 and 3. Let T' denote a CT, whose nodes components are nodes 4, 5, and 6. The KC is T. KSRQuerying considers $T' \in IR_T$, since T' does not satisfy properties 1, 2, and/or 3. Therefore, it returned image node $6 \in T'$ as the answer.

- Consider Fig. 12-B, which presents information about a conference and its collocated workshops. Nodes 4 and 7 contain the subject titles of the conference and one of its workshops. Now consider the query Q (name ="ICDE", subjTitle?). The query asks for the subject title of the ICDE conference (node 2). The correct answer is node 4. But, [13] returned both nodes 4 and 7, because the LCA of each of them with node 2 is the same node (node 1).

KSRQuerying answer: Let *T* denote a CT, whose nodes components are nodes 1, 2, 3, and 4. Let *T'* denote a CT, whose nodes components are nodes 5, 6, and 7. The KC is *T*. Since *T* and *T'* have the same Ontology Label, $T' \notin IR_T$ (recall property 1). Therefore, KSRQuerying returned only the subjTitle node $4 \in T$.

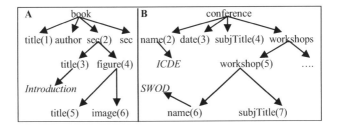

Fig. 12. Fragments of XML documents taken from [11]

9.2 Search Performance Evaluation

To evaluate the query execution times of KSRQuerying under different document sizes, we ran all the queries of XMark [22] using documents of variable sizes (200, 250, and 300 MBs). For *each* of the four document sizes, we ran all the 20 queries of XMark and computed the *average* query execution time of KSRQuerying, [13], [14], and [21]. For the sake of fair performance comparison with the other systems, we first used *each* system S_i to *precompute* the relationships between all nodes in all documents (before queries are submitted to S_i), saved the results for future accesses by S_i, and recorded the computation time "t". We considered "t" as constant for S_i: avg query execution time of S_i = (t + execution time of all queries)/number of queries. For SRQuerying, "t" included the time for creating a CTG, IRs, and Ontology Labels (but it did not include the time of building taxonomies of concepts for the previously mentioned 25% of the tag names). Figure 13 shows the results. As can be seen, the average query execution time of KSRQuerying is less than those of Schema-Free XQuery and XSeek, and it is slightly higher than the average query execution time of XKSearch. The slight performance of XKSearch over KSRQuerying is due to the

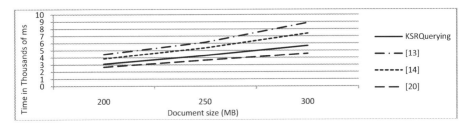

Fig. 13. Execution times of KSRQuerying, [13, 14, 21] on XMark using variable document sizes

overhead of applying the context-driven search techniques. The performance of KSRQuerying over Schema-Free XQuery [13] and XSeek [14] is due to: (1) KSRQuerying's recursive querying capability, (2) the computation overhead of XSeek's inference mechanism for determining *desirable* results nodes, and (3) [13] builds a relationship between *each* two nodes containing keywords, and *then* filter results according to the search terms.

10 Conclusions

We proposed an XML search engine called KSRQuerying, which answers recursive queries, keyword-based queries, and loosely structured queries. We experimentally evaluated the quality and efficiency of KSRQuerying and compared it with the systems proposed in [13, 14, 21]. The results showed that the recall and precision of KSRQuerying outperform those of [13, 14, 21]. The tests results showed also that the average query execution time of KSRQuerying is less than those of [13, 14] and is slightly higher than [21]. KSRQuerying created an Ontology Label for each *distinct* tag name in INEX [11, 12] and XMark [22]. We used Protégé for creating about 25% of the ontologies and the other 75% were available in electronic form (done by others) and we imported them to the system using namespaces co-ordination.

References

1. Alorescu, D., Manolescu, I.: Integrating Keyword Search into XML Query Processing. Computer Networks 33, 119–135 (2002)
2. Agrawal, C., Das, G.: DBXplorer: a System for Keyword-Based Search Over Relational Databases. In: ICDE 2002 (2002)
3. Aditya, B., Sudarshan, S.: BANKS: Browsing and keyword Searching in Relational Databases. In: VLDB 2002 (2002)
4. Balmin, A., Koudas, N.: A System for Keyword Proximity Search on XML Databases. In: VLDB 2003 (2003)
5. Balmin, A., Hristidis, V., Papakonstantinon, Y.: Keyword Proximity Search on XML Graphs. In: ICDE 2003 (2003)
6. Balmin, A., Hristidis, V.: ObjectRank: Authority-Based Keyword Search in Databases. In: VLDB 2004 (2004)
7. Botev, C., Shao, F.: XRANK: Ranked Keyword Search over XML Documents. In: SIGMOD 2003 (2003)
8. Cohen, S., Kanza, Y.: Interconnection Semantics for Keyword Search in XML. In: CIKM 2005 (2005)
9. Hristidis, V., Papakonstantinou, Y.: DISCOVER: Keyword search in Relational Databases. In: VLDB 2002 (2002)
10. Initiative for the Evaluation of XML Retrieval (INEX) (2005), http://inex.is.informatik.uni-duisburg.de/2005/
11. Initiative for the Evaluation of XML Retrieval (INEX) (2006), http://inex.is.informatik.uni-duisburg.de/2006/
12. Jagadish, H., Li, Y., Cong, Y.: Schema-Free XQuery. In: Proc. VLDB 2004 (2004)
13. Liu, Z., Chen, Y.: Identifying Meaningful Return Information for XML Keyword Search. In: SIGMOD 2007 (2007)

14. List of Ontology editor tools,
    http://www.xml.com/2002/11/06/Ontology_Editor_Survey.html
15. Protégé ontology editor, http://protege.stanford.edu/
16. Taha, K., Elmasri, R.: OOXSearch: A Search Engine for Answering Loosely Structured XML Queries Using OO Programming. In: Cooper, R., Kennedy, J. (eds.) BNCOD 2007. LNCS, vol. 4587, pp. 82–100. Springer, Heidelberg (2007)
17. Taha, K., Elmasri, R.: CXLEngine: A Comprehensive XML Loosely Structured Search Engine. In: Proc. DataX 2008 (2008)
18. ToXgene, a template-based generator for large XML documents,
    http://www.cs.toronto.edu/tox/toxgene/
19. TIMBER, http://www.eecs.umich.edu/db/timber/
20. Xu, Y., Papakonstantinou, Y.: Efficient Keyword Search for Smallest LCAs in XML Databases. In: SIGMOD 2005 (2005)
21. XMark — An XML Benchmark Project,
    http://monetdb.cwi.nl/xml/downloads.html
22. XQEngine: downloaded from,
    http://sourceforge.net/projects/xqengine/

The XML-λ XPath Processor: Benchmarking and Results

Jan Stoklasa and Pavel Loupal

Dept. of Computer Science and Engineering
Faculty of Electrical Engineering, Czech Technical University
Karlovo nám. 13, 121 35 Praha 2
Czech Republic
stoklj2@fel.cvut.cz, loupalp@fel.cvut.cz

Abstract. This paper presents XML-λ, our approach to XML processing based on the simply typed λ-calculus. A λ-calculus model of both the XML and XPath languages is described and a prototype implementation is investigated.

We benchmark the prototype implementation, comparing it to existing XPath processors — Apache Xalan, Saxon, and Sun's Java JAXP. Surprisingly, although the prototype is more of an idea validation tool than a benchmark tuned software, XPath query evaluation is fast, especially on pre-loaded XML documents. Benchmark results support our decision to use XML-λ as an XML database data storage model.

This work is part of a long-term effort targeted at designing a native XML database management system built upon this theoretical model.

1 Introduction

XPath [2] is a fundamental W3C specification for addressing parts of an XML document. It is used as a basic part in many other W3C specifications such as XLink, XPointer, XSLT or XQuery.

Here, we publish our contribution to evaluate XPath expressions using a λ-calculus based framework called XML-λ. This framework proposes a way of modeling XML schema by a set of functions and also includes a definition of a query language based on the simply typed λ-calculus. Apart from its formal specification, we already have a working prototype of an XPath processor. This paper is devoted to comparing XML-λ performance with that of Apache Xalan [17], Saxon [6], and Sun's Java JAXP (the one available in Sun's Java SE Runtime Environment version 1.6) [16].

Contributions. The achievements depicted in this paper are results of our long-running endeavor to propose and implement a functional approach for querying and processing XML data. This work is a logical step forward to practical verification of the proposal.

Z. Bellahsène et al. (Eds.): XSym 2009, LNCS 5679, pp. 53–66, 2009.

The main contributions of this article are the following:

- We prototype and examine a theoretically sound[1] approach to XPath query evaluation.
- We describe a prototype implementation of the XML-λ theoretical model written in Java.
- We publish results of performance benchmarking the XML-λ prototype and state-of-the-art XPath processors showing that the functional model performs well in practice.
- We show that XML-λ is an appropriate model for data storage in an XML database management system.

Structure. The rest of this paper is structured as follows: Section 2 lists related projects and works. In Section 3 we briefly repeat basic facts about the XML-λ Framework with links to more detailed description. Further, a short overview of the processor implementation follows in Section 4. Main parts of this work, specification of the benchmark and benchmark results are shown in Sections 5 and 6, respectively. Within Section 7 we discuss our results and then conclude with outlook to future work in Section 8.

The complete list of XPath queries and corresponding results of the performance benchmark is available at [15]. The subset of queries actually used in the benchmark along with relevant results is listed in Appendix A and Appendix B, respectively.

2   Related Work

Since we are not familiar with any similar work (i.e. benchmarking of an XPath implementation based on λ-calculus model), in this section we deal with related topics: XPath semantics, XPath processors, XQuery processors, and XML benchmarking.

XPath semantics. The World Wide Web Consortium, the originator of the XPath language, published a semi-formal semantics of XPath and XQuery [3]. Wadler defined a denotational semantics for the XSLT pattern language [18] (this language was consequently introduced as XPath) and proposed a set-based model for it using this sort of semantics.

XPath processors. Usually, XPath processors are part of XSLT processors and do not exist as standalone products. On the other hand, every XSLT processor is able to execute ad-hoc XPath queries. Popular XPath/XSLT processors are Saxon [6] and Apache Xalan [17].

XQuery processors. The XML Query Language (XQuery) is the most popular language for XML nowadays with many existing implementations; the most mature are Galax [4] or eXist [10]. Indeed, our long-term effort is also targeted at designing a native XML database that uses XQuery as a query language based on the XML-λ as an XQuery implementation model.

---

[1] The XML-λ soundness proof is part of a dissertation thesis to be submitted at the Czech Technical University in Prague.

XML benchmarking. Franceschet designed XPathMark [5], a set of tests evaluating both the correctness and performance of an XPath processor implementation - the XML-λ benchmark described in this paper uses XPathMark.

The XMark Benchmark [14] is a toolkit for evaluating the performance of XML databases and query processors. The `xmlgen`, a valuable tool generating well-formed XML files according to a given XML Schema, is part of the toolkit and is extensively used in the XML-λ benchmark.

The Kawa language framework [1] is a Java framework aiming to implement dynamic languages on top of the Java Virtual Machine. A Scheme programming language dialect called Kawa was implemented using the Kawa language framework and there is also a partial XQuery implementation called Qexo. Comparing XML-λ to Kawa, we conclude that the Kawa language framework is an inspiring effort indeed, investigating XML query languages from the point of view of dynamic programming languages implementation. XML-λ model stays closer to the XML world, investigating XML query languages from the XML database point of view.

3   The XML-λ Framework

XML-λ is a functional framework for processing XML. The original proposal [12,13] defines its formal base and shows its usage primarily as a query language for XML but there is also a consecutive work that introduces updates into the language available in [8].

3.1   Concept and Basic Definitions

In XML-λ there are three important components related to its type system: *element types*, *element objects*, and *abstract elements*. We can imagine these components as the data dictionary in relational database systems. Note also Figure 1 for relationships between basic terms of W3C standards and the XML-λ Framework.

Element types are derived from a particular Document Type Definition (DTD). For each element defined in the DTD there exists exactly one *element type* in the set of all available element types (called T_E). Consequently, we denote E as a set of *abstract elements*.

Element objects are basically functions of type either $E \rightarrow String$ or $E \rightarrow (E \times \ldots \times E)$. Application of these functions to an *abstract element* allows access to element's content. *Elements* are, informally, values of *element objects*, i.e. of functions. For each $t \in T_E$ there exists a corresponding t-object (an element object of type t).

Finally, we can say that in XML-λ the instance of an arbitrary XML document is represented by a subset of E and a set of respective t-objects.

3.2   Example

The following example describes the way that the type system T_E (both element types and functional types) is constructed and briefly explains how we access

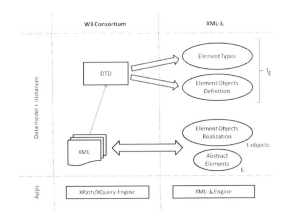

Fig. 1. The relationship between W3C and XML-λ models

an XML instance. Note that the DTD we present in Figure 2 is the one we use later in the benchmark.

```
<!ELEMENT site
    (regions, categories, catgraph, people,
    open_auctions, closed_auctions)>
<!ELEMENT categories  (category+)>
<!ELEMENT category    (name, description)>
<!ELEMENT name        (#PCDATA)>
<!ELEMENT description (text | parlist)>
<!ELEMENT text
    (#PCDATA | bold | keyword | emph)*>
<!ELEMENT keyword
    (#PCDATA | bold | keyword | emph)*>
<!ELEMENT parlist       (listitem)*>
<!ELEMENT listitem    (text | parlist)*>
```

Fig. 2. An example DTD

For given schema we construct the set of element types as

$$SITE : (REGIONS, CATEGORIES, CATGRAPH,$$
$$\qquad PEOPLE, OPEN\_AUCTIONS,$$
$$\qquad CLOSED\_AUCTIONS) >,$$
$$CATEGORIES : CATEGORY+,$$
$$CATEGORY : (NAME, DESCRIPTION),$$
$$NAME : String,$$
$$DESCRIPTION : (TEXT \mid PARLIST),$$
$$TEXT : (String \mid BOLD \mid KEYWORD \mid EMPH)*,$$
$$KEYWORD :,$$

$(String \mid BOLD \mid KEYWORD \mid EMPH)*,$
$PARLIST : LISTITEM*,$
$LISTITEM : (TEXT \mid PARLIST)*.$

Further, we define functional types (denoted as t-objects) as

$SITE : E \rightarrow (E \times E \times E \times E \times E \times E),$
$CATEGORIES : E \rightarrow 2^E,$
$CATEGORY : E \rightarrow (E \times E),$
$NAME : E \rightarrow String,$
$DESCRIPTION : E \rightarrow E,$
$TEXT : E \rightarrow 2^E,$
$KEYWORD : E \rightarrow 2^E,$
$PARLIST : E \rightarrow 2^E,$
$LISTITEM : E \rightarrow 2^E.$

Having looked at Figure 3, showing a fragment of an XML document valid according to the `auction.dtd` (the original DTD used in the XPathMark project [5]), we can see that there are thirteen abstract elements (members of E).

```
<?xml version="1.0" standalone="yes"?>
<site>
  <regions>
    <africa>
      <item id="item0">
        <location>United States</location>
        <quantity>1</quantity>
        <name>duty</name>
        <payment>Creditcard</payment>
        <description>
        <parlist>
          <listitem>
            <text>page rous lady
              <keyword>officer</keyword>
            </text>
          </listitem>
            ...
</site>
```

Fig. 3. Fragment of a valid XML instance document

In this scenario, for instance, the *name*-element object (a function of type $E \rightarrow String$) is defined exactly for one abstract element (the one obtained from the `<name>duty</name>` XML element) and for this abstract element it returns value "duty".

As a more complex example, function *item* of type $ITEM : E \rightarrow (E \times E \times E \times E \times E \times E)$, applied to the `<item>` element, returns a six-tuple — subset of this

Cartesian product. From the tuple we can then obtain its particular components by performing name-based projections (through element type names).

3.3   Query Language

A typical query has an expression part — a predicate to be evaluated over data — and a constructor part that wraps query result and forms the XML output. We retain this model as well together with the fact that the expression part in the XML-λ query language is based on λ-terms defined over the type system T_E.

Main constructs of the language are variables, constants, tuples, use of projections and λ-calculus operations — applications and abstractions. Syntax of this language is similar to λ expression, i.e. $\lambda \ldots (\lambda \ldots (expression) \ldots) \ldots$ In addition, there are also typical constructs such as logical connectives, constants or relational predicates.

Due to paper length constraints we will not discuss principles of the language in detail here (these can be found in [12] or [13]). As an example for the construction of XML-λ queries, let us consider a query that returns all items from a specified location. In XPath, we write this query as

```
//item[location = "United States"]
```

By an automated translation that is described in our previous work [9] we obtain equivalent XML-λ query as follows

```
xmldata("auction.xml")
lambda x (/item(x) and
          x/location="United States"))
```

3.4   Summary

There are still open issues related to the XML-λ Framework. Perhaps the main drawback of our prototype implementation is the lack of optimizations and limited support for indexes – the only type of index supported is a parent-child relationship index. Here, we see good research opportunities in (a) designing additional indexes and (b) optimizing the XML-λ evaluation using functional programming techniques such as lazy evaluation.

Notwithstanding, we already have a suitable software library to be embedded into the ExDB database management system on which we are working [7]. Its current implementation uses a basic persistent DOM storage but in the near future we plan to replace it with the XML-λ Framework core libraries.

4   Processor Implementation

Our primary goal was to keep XML-λ prototype design as close to the theoretical model as possible. It is a natural requirement with respect to the fact that this prototype is a proof-of-concept implementation of the XML-λ Framework. Due to space restriction, we just highlight the most important facts.

XML documents are parsed by a SAX parser and turned into an in-memory XML-λ model. Each document is realized as

1. a HashMap[2] of element types,
2. an unordered set of AbstractElements,
3. two HashMaps of PCData and CData items, and
4. a HashMap of t-objects mapping parent-child relationships and a HashMap storing inter-type relationships.

XPath queries are parsed using the ANTLR LL(k) tool [11]. The resulting Abstract Syntax Tree (AST) is processed using the Visitor design pattern utilizing a recursive descent parser. AST subtrees are evaluated and the value obtained in the root of the AST is returned as the evaluation result.

The current version of the XPath processor does not support the complete XPath 1.0 specification yet; notably the id() function and some of the navigation axes are missing. Therefore we chose only a subset of queries for the experiment.

5 Benchmark Environment

The xmlgen tool, developed at CWI as a part of the XML Benchmark Project [14], was used to generate a set of input XML files. Setting the XMark factor parameter values to 0.01, 0.02, 0.05, 0.1 and 0.2 in successive steps, the test set containing 1.12 MB, 2.27 MB, 5.6 MB, 11.3 MB and 22.8 MB XML files was obtained.

XPath queries used in this benchmark are a subset of queries used by the XPath Performance Test [5]. The selection of XPath Performance Test queries was determined by features supported by the XML-λ prototype implementation. The current XML-λ prototype is able to correctly process 21 XPath Performance Test queries: A1-A8, B1-B10, C2, C3, and E5.

Following *XPath processors* were tested: XML-λ Sun's Java JAXP 1.4, Saxon 9.1.0.2, and Apache Xalan 2.7.1. XPath processors were deployed as a set of JAR files and a test harness program was written for each XPath processor. XML parsing, XPath query compilation and XPath query evaluation times as well as total time were measured.[3]

Shell scripts ran the XPath test harness programs in sequence increasing the XML input file size. Result correctness was verified using the diff tool. The Scheme programming language was used as a kind of metalanguage generating shell script for each of the benchmark runs.

The benchmark was run on AMD Athlon X2 1.90 GHZ PC with 1 GB RAM running Java Runtime Environment 6 on Windows XP. All the daemon programs

---

[2] A HashMap data structure is implemented as a hash table in the Java library, guaranteeing the $O(1)$ amortized time complexity of an element object access.

[3] XPath processors create indexes and auxiliary data structures during XML parsing phase, therefore it makes sense to measure individual processing phases separately and a total time as well.

(called services in Windows environment) were shut down to minimize external interference and swap file usage was disabled.

In this benchmark, we were interested in time complexity of XPath processor implementations, leaving space complexity and memory usage aside. However, we monitored the memory usage of running XPath processor implementations using operating system log and we plan to measure memory usage using the Java jstat tool.

6   Results

All measured results are available on the web [15] as stated earlier. Now, let us investigate in detail the C3 query:

```
/site/people/person[profile/@income =
/site/open_auctions/open_auction/current]
/name
```

We can see the result of C3 time measurement in Figure 4.

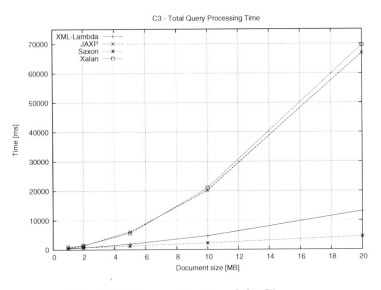

Fig. 4. Total processing time of the C3 query

Without doubt, the best results are gained by the Saxon XPath processor. It performs best for most queries and scales well for different document sizes. On the other hand, none of the examined XML processors outperformed all the competitors in all cases – e.g., Java JAXP is the fastest one processing small instances of A1 query, but its results downgrade for larger instances.

An interesting detail to note is almost exact ratio 2:1 of Java JAXP results in tests B9 and B10. This can be explained looking at B9 and B10 queries, with B9

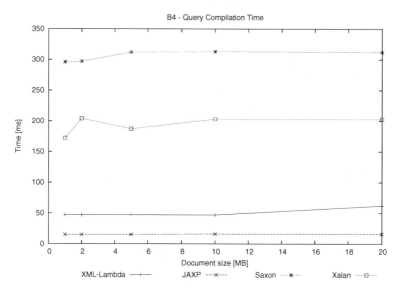

Fig. 5. B4 – query compilation time

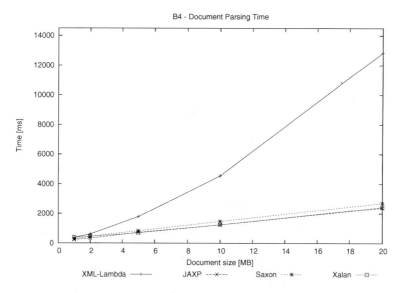

Fig. 6. B4 – document parsing time

using the or operator and B10 using **and** operator in the XPath filter clause. On a typical XML file used here, the first operand (`not(bidder/following::bidder)` `or not(bidder/preceding::bidder)`) evaluates to false in most cases, since almost all auctions have more than one bidder. So it follows that in processing

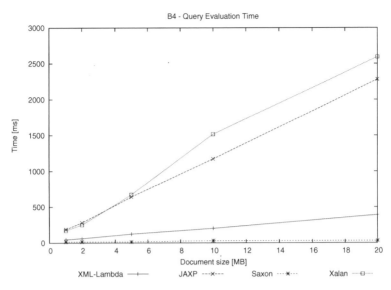

Fig. 7. B4 – query evaluation time

of the B9 query Java JAXP has to evaluate the second operand as well, but in case of processing the B10 query the second operand can be ignored.

A detailed testcase. As a typical example, let us have a look at Figures 5, 6, and 7. These plots show results for the B4 query but such a behavior is typical for most of the queries in the benchmark.

Note that the XML-λ document parsing performance lacks compared to DOM, however the query compilation and query evaluation times are promising.

7   Discussion

Examining *small document instances*, the XML-λ processor slightly outperforms Saxon and Xalan for most queries. For *larger instances* XML-λ document parsing performance downgrades, but the query evaluation time is still excellent for most queries.

Weak results gained by XML-λ for queries B5, B6, B9, B10 and E5 (and also those for B9 and B10 achieved by Java JAXP and Xalan) are obviously caused by lack of optimizations inside these processors. Detection of tautologies and contradictions would improve these results significantly.

Naturally, for XML-λ we know the intimate reason for such behavior. Its data model is based on sets and evaluating parent-child relationships and converting sets to lists (and vice-versa) consumes plenty of time. The parent-child index helps here, but we still plan to do more optimizations in the future.

We believe this once more stresses importance of a clear theoretical model behind the XPath processor implementation. Ad-hoc implementations may experience unexpectedly low performance for some queries and it is hard to reason about optimizations without formal semantics at hand.

8   Conclusions

We have performed the XPathMark performance benchmark on three state-of-the-art XPath processors implemented in Java and compared the results with our XML-λ prototype. With respect to the fact that the prototype is in an early stage of development, we did not expect auspicious results. Despite it, the results are undoubtedly comparable in most cases.

There are three main outcomes of this experiment related to the XML-λ processor: (1) The benchmark has not found any serious functional errors in the prototype. (2) For small document instances the XML-λ is the second best implementation and for large instances it lacks on document parsing time but performs well on query evaluation time. There is an important fact that all these results were achieved without optimizations in the source code – the XML-λ implementation in Java is isomorphic to the XML-λ formal semantics definition without any performance tweaks in the source code. (3) Measuring just the time spent evaluating the query (ignoring the query compilation and the XML parsing phase), XML-λ performance is excellent. On the other hand, parsing an XML document into the XML-λ model is 2-3 times slower than parsing an XML document using DOM. These results support our claim that XML-λ is an appropriate model for the data storage in the XML database where the XML document parsing usually happens once followed by multiple queries being processed.

Future Work. The aim of this submission was a preliminary check whether the concept of the functional framework has a chance to survive in the quickly evolving domain of XML. We can express our satisfaction with the results but there is still a lot of work ahead. From the theoretical point of view, we plan to extend our approach to XQuery. The fact that XPath 2.0 is its subset helps us much but there are still open topics, namely FLWOR expressions and output construction.

Simultaneously, we will improve the prototype implementation because there is a lot of work to be done on various optimization methods.

References

1. Bothner, P.: The Kawa language framework,
   http://www.gnu.org/software/kawa/
2. Clark, J., DeRose, S.: XML Path Language (XPath) 1.0 (November 1999),
   http://www.w3.org/TR/xpath
3. Draper, D., Fankhauser, P., Fernández, M., Malhotra, A., Rose, K., Rys, M., Siméon, J., Wadler, P.: XQuery 1.0 and XPath 2.0 Formal Semantics (January 2007), http://www.w3.org/TR/xquery-semantics/

4. Fernández, M., Siméon, J.: Galax (2004),
   http://www-db-out.bell-labs.com/galax/
5. Franceschet, M.: XPathMark: An XPath Benchmark for the XMark Generated
   Data. In: Bressan, S., Ceri, S., Hunt, E., Ives, Z.G., Bellahsène, Z., Rys, M., Unland,
   R. (eds.) XSym 2005. LNCS, vol. 3671, pp. 129–143. Springer, Heidelberg (2005)
6. Kay, M.: Saxon - XSLT Transformer (2001), http://saxon.sourceforge.net/
7. Loupal, P.: Experimental DataBase (ExDB) Project Homepage,
   http://swing.felk.cvut.cz/~loupalp
8. Loupal, P.: Updating typed XML documents using a functional data model. In:
   Pokorný, J., Snášel, V., Richta, K. (eds.) DATESO. CEUR Workshop Proceedings,
   vol. 235, CEUR-WS.org (2007)
9. Loupal, P., Richta, K.: Evaluation of XPath Fragments Using Lambda Calculi.
   In: ITAT 2008 - Information Technologies - Applications and Theory, pp. 1–4.
   Univerzita P.J.Šafárika, Košice (2008)
10. Meier, W.: eXist, http://exist.sourceforge.net/
11. Parr, T.: The Definitive ANTLR Reference: Building Domain-Specific Languages.
    The Pragmatic Bookshelf (2007) ISBN: 9780978739256
12. Pokorný, J.: XML functionally. In: Desai, B.C., Kioki, Y., Toyama, M. (eds.) Pro-
    ceedings of IDEAS2000, pp. 266–274. IEEE Comp. Society, Los Alamitos (2000)
13. Pokorný, J.: XML-λ: an extendible framework for manipulating XML data. In:
    Proceedings of BIS 2002, Poznan, pp. 160–168 (2002)
14. Schmidt, A.R., Waas, F., Kersten, M.L., Florescu, D., Manolescu, I., Carey, M.J.,
    Busse, R.: The XML Benchmark Project (April 2001)
15. Stoklasa, J., Loupal, P.: Complete Results of the XML-λ Benchmark,
    http://f.lisp.cz/XmlLambda
16. Sun Microsystems, Inc.: Java API for XML Processing (JAXP) (2006),
    https://jaxp.dev.java.net/
17. The Apache Software Foundation: Apache Xalan - XSLT Transformer (2001),
    http://xml.apache.org/
18. Wadler, P.: A formal semantics of patterns in XSLT. In: Markup Technologies,
    pp. 183–202. MIT Press, Cambridge (1999)

A   List of Queries

Table 1. Selected queries from the XPathMark benchmark

A1	`/site/closed_auctions/closed_auction/annotation/description/text/keyword`
A2	`//closed_auction//keyword`
A3	`/site/closed_auctions/closed_auction//keyword`
A4	`/site/closed_auctions/closed_auction[annotation/description/text/keyword]/date`
A5	`/site/closed_auctions/closed_auction[descendant::keyword]/date`
A6	`/site/people/person[profile/gender and profile/age]/name`
A7	`/site/people/person[phone or homepage]/name`
A8	`/site/people/person[address and (phone or homepage) and (creditcard or profile)]/name`
B1	`/site/regions/*/item[parent::namerica or parent::samerica]/name`
B2	`//keyword/ancestor::listitem/text/keyword`
B3	`/site/open_auctions/open_auction/bidder[following-sibling::bidder]`
B4	`/site/open_auctions/open_auction/bidder[preceding-sibling::bidder]`
B5	`/site/regions/*/item[following::item]/name`
B6	`/site/regions/*/item[preceding::item]/name`
B7	`//person[profile/@income]/name`
B8	`/site/open_auctions/open_auction[bidder and not(bidder/preceding-sibling::bidder)]/interval`

B9	`/site/open_auctions/open_auction[(not(bidder/following::bidder) or`
	`not(bidder/preceding::bidder)) or`
	`(bidder/following::bidder and bidder/preceding::bidder)]/interval`

B10	`/site/open_auctions/open_auction[(not(bidder/following::bidder) or`
	`not(bidder/preceding::bidder)) and`
	`(bidder/following::bidder and bidder/preceding::bidder)]/interval`

C2	`/site/open_auctions/open_auction[bidder/increase = current]/interval`
C3	`/site/people/person[profile/@income = /site/open_auctions/open_auction/current]/name`
E5	`/site/regions/*/item[preceding::item[100] and following::item[100]]/name`

B   Benchmarking Results

Table 2. Total processing time for the C3 query [ms]

Document size [MB]	1	2	5	10	20
Saxon	641	766	1.266	2.234	4.469
XML-λ	453	703	1.906	4.672	13.126
Java XPath	609	1.390	6.110	20.173	66.910
Xalan	922	1.485	5.594	21.017	69.535

Table 3. Total processing time for the B9 and B10 queries for the Java XPath [ms]

Document size [MB]	1	2	5	10	20
B9	30.314	106.897	1.340.139	9.681.495	70.740.387
B10	15.000	53.768	642.756	4.858.452	35.399.742

Table 4. Complete results for the B4 query

Document size [MB]		1	2	5	10	20
XML-λ	Query compilation [ms]	46	47	47	46	47
	Document parsing [ms]	375	609	1.765	4.516	12.814
	Query evaluation [ms]	47	63	110	204	375
	Total time [ms]	**468**	**719**	**1.922**	**4.766**	**13.236**
Java XPath	Query compilation [ms]	15	16	16	0	16
	Document parsing [ms]	219	313	719	1.235	2.344
	Query evaluation [ms]	188	265	625	1.172	2.297
	Total time [ms]	**422**	**594**	**1.360**	**2.407**	**4.657**
Saxon	Query compilation [ms]	297	297	297	297	313
	Document parsing [ms]	281	406	797	1.484	2.672
	Query evaluation [ms]	16	16	16	32	47
	Total time [ms]	**594**	**719**	**1.110**	**1.813**	**3.032**
Xalan	Query compilation [ms]	188	187	187	203	203
	Document parsing [ms]	359	438	688	1.203	2.406
	Query evaluation [ms]	156	266	672	1.500	2.594
	Total time [ms]	**703**	**891**	**1.547**	**2.906**	**5.203**
Number of nodes in the result		602	928	2.446	5.102	9.955

XPath+: A Tool for Linked XML Documents Navigation

Paulo Caetano da Silva and Valéria Cesário Times

Federal University of Pernambuco, Center for Informatics, Brazil, P.O. BOX 7851
paulo.caetano@bcb.gov.br, vct@cin.ufpe.br

Abstract. Links are basic elements in the World Wide Web. The use of links in XML documents goes further than in the WWW, since XML links express the semantics of a relationship. XLink has been proposed by the W3C as a standard for representing links in XML. However, most of the current query languages found in literature, like XPath, do not support navigation over XML links, making its adoption difficult by software developers. In this paper, an extension for the XPath query language is proposed, namely XPath+, which provides a means of navigating through both internal and external links. Particularly, both the syntax and semantics of XPath+ are given, along with some results derived from the implementation of our work.

Keywords: XPath+, XLink, XML, XPath, XBRL.

1 Introduction

As a result of the increasing need of data integration and data exchange, XML documents are turning into very large and interlinked files. Often, these documents have complex link networks pointing to all kinds of resources. These resources should be used only combined with the XML document, since their semantics are defined by the link networks. The XML Linking Language (XLink) [1] is used to describe relationships among resources included in XML documents by links. Processing documents with link networks has become a challenging task, because query languages do not support link traversing techniques. XPath [2] has been seen as a de facto standard in the XML query research area. However, it does not provide a means of navigating through XLink links. As a result, both the semantics and the processing issues concerning link data are compromised. By the way of example, XBRL (eXtensible Business Reporting Language) [3] is a XML-based language used to create business reports, mostly adopted by the financial field. It uses plenty of XLink links in order to express the semantics of instance elements. XBRL links establish associations among business concepts and between concepts and the document. This structural characteristic raises a strong need for XLink processors.

In this paper, we present the language XPath+, which is an extension of the XPath language. It fully supports link traversing in XML documents, whether the link is inside the document or in a separate linkbase. A link is used to associate two or more resources, as distinct from a linkbase, which is a separate XML document that holds a set of links. Our goal is to develop a system compliant with widely adopted technologies (XPath, Java and DOM), which can be used in a context of heterogeneous data

Z. Bellahsène et al. (Eds.): XSym 2009, LNCS 5679, pp. 67–74, 2009.

sources, allowing to query linked XML documents. This paper is organized as follows. Section 2 presents our contribution, the XPath+ language, including its syntax and properties, together with some issues related to its application and query processor. Following this, section 3 discusses the usefulness of the XPath+ language in a practical scenario, based on financial data represented in XBRL documents. Next, section 4 presents a comparative analysis between XPath+ and some related work. Finally, section 5 summarizes the proposed work and identifies some of the next tasks of our research.

2   XPath+

This section presents a language that supports navigation over all links defined in the XLink. For this, it was necessary to extend the XPath with new functions and axes. XLink links can be found in an instance document or in a schema [4] or grouped in a linkbase, by the use of the *schemaLocation* and *xlink:href* attributes. For all possible link locations, see Figure 1. The main concepts added to XPath are the new axes *link-source* and *link-destination*, whose semantics is similar to the XPath axes *child* and *parent*. Using these axes, it is possible to navigate across resources, whether they are local or remote elements.

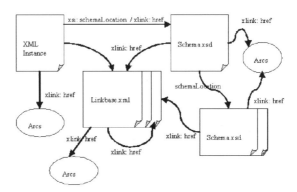

Fig. 1. XML Link Document Chain

2.1   A Data Model for XML with XLink

A data model based on XML documents and a binary axis relations were defined by *Gottlob et al* [5] to discuss an algorithm which evaluates XPath expressions and improves the efficiency of queries regarding the time and space. Motivated by applications to XML, *Libkin* [6] examined query languages for unranked trees and presented a set of definitions to handle XML data. However, these results do not deal with links. In order to provide navigation over links, we have extended their results and thus, the following definitions were obtained.

Definition 1. A labeled unranked rooted tree is a tree with no bound on the number of children, in which each node is given a unique label s that is an element of N^{*}

(i.e. finite strings of natural numbers). We then define a labeled unranked ordered rooted tree T as a pair (D, $<_{pre}$) where: (1) the string s (the empty string) is the root; (2) D, the set of nodes called tree domain, is a prefix-closed finite subset D of N^* such that s ∈ D implies w ∈ D if and only if w $<_{pre}$ s. The relation $<_{pre}$ is the prefix relation on the elements of D, such that w $<_{pre}$ s, if and only if the unique path from the root to s passes through w. An XML document is a data structure defined as follows:

Definition 2. A XML document d is represented as the five-tuple (T, β, λ, Rχ, Rσ) where: (1) T= (D, $<_{pre}$) is a labeled unranked ordered rooted tree; (2) β is the set of tags; (3) λ: D → β is a function which assigns one node in T to each XML tag; (4) Rχ is a set of binary axis relations over β; (5) Rσ is a set of binary relations over β′× β″, where β′ and β″ are sets of tags of documents d′ and d″, respectively, with d′≠ d″.

Definition 3 (axis function). Let d be a XML document (T, β, λ, Rχ, Rσ) . For an *XPath+* axis relation χ ⊆ Rχ we define a function f_χ : ρ(β)→ P(β) (and thus overload the relation name χ, as for example f_{child}), where ρ(β) is the power set of β, as f_χ(X) = {y ∈β I ∃x ∈ X such that (x,y) ∈ χ}.

   In *XPath+*, besides the relationship among nodes in a given XML document we also define the relationship among nodes of two distinct XML documents:

Definition 4 (σ link function). Let d′ and d″ be documents (T′, β′, λ′, Rχ′, Rσ′) and (T″, β″, λ″, Rχ″, Rσ″), respectively. For an *XPath+* relation σ ⊆ Rσ′ in d′ we define a function g_σ : P(β′)→ P(β″) as g_σ(X) ={ y ∈β″ I ∃x ∈ X such that (x,y) ∈ σ}.

   From the given definitions, it is possible to define the function that navigates over XPath+ expressions. Let φ be a query expression, it can be expressed as φ = {(σ(x_i) Iχ$_{(i)}$)}, where x_i is an element defined in the hierarchy, and the query may be executed on the hierarchy or on links. The *NavigationPath* function executes the queries and is defined by the algorithm given in Figure 2. Its arguments are a query and a node list that contains the elements of an XML document. The function evaluates the parameters and defines if the query can be solved by an XPath processor. If this is not the case and there are links in the query, then it is processed by an XPath+ processor, using the *xpath+Evaluator* function, defined by the algorithm shown in Figure 2. The *xpath+Evaluator* function receives as input parameters an XPath+ query and a node list. Following this, it goes through the node list looking for arcs. When a reference to a linkbase is found, the function checks if it references other arcs or linkbases and then includes them in the search. Then, the function executes the same process on schemas by considering all the link access ways.

2.2 Functions

In order to facilitate user navigation over elements that contain links, some functions were added to the XPath specification. These functions are: (1) Function *fn:isLinkSource($arg as item()*) as item()**: returns a Boolean value that indicates if the context node is a link source, that is, if there are any links whose *from* attribute points to the context element; (2) Function *fn:isLinkDestination($arg as item()*) as item()**: returns a Boolean value that indicates if the context node is a link destination, that is, if there are any links whose *to* attribute points to the context element. The XPath+ version of some XPath functions are slightly different. The functions *last()*, *position()* and *count()* must work also on node lists created by links. For example, the *link-destination::* axis is used in an expression and its result is a node list with five elements. Consider that the function *count( )* is then applied to the node list. As a result, the value five should be returned. If, instead, the function *last( )* is applied to the same list, the result should be the fifth element of that list. Finally, if the function *position( )* is applied, it should return the current position of a node in the list.

```
φ = {(σ(x) |χ()}
function NavigatePath (φ, listNode)
begin
    if φ is xpathExpression then
        return xptahEvaluator(φ, listNode)
    else
        return xptah+Evaluator(φ, listNode)
    endif
end
```

```
function xpath+Evaluator(φ, listNode)
begin
    search instance arcs
    for each 0 ≤ j ≤ n in instance do
        search linkbase [j] arcs
        for each 0 ≤ k ≤ n in linkbase [j] do
            search linkbase [k] arcs
        end for
    end for
    for each 0 ≤ j ≤ n in instance do
        search schema [j] arcs
        for each 0 ≤ k ≤ n in schema [j] do
            search linkbase [k] arcs
            for each 0 ≤ i ≤ n linkbase [k] do
                search linkbase [i] arcs
            end for
        end for
    end for
    validate(arcs, instance)
    return arcs
end
```

Fig. 2. NavigationPath and xpath+Evaluator Algorithms

2.3 Syntax and Semantics

The *link-source::* axis selects a list of nodes having the source node as the context node. In Figure 3, to find the elements that the *bcb:currentAssets* points to, the XPath+ expression is */link-source:: bcb:currentAssets*. Likewise, the *link-destination::* axis selects a list of nodes whose links point to the context node. If one needs to find the children of the *assets* element, the XPath+ expression is *"/link-destination::bcb:assets"*. Its result is shown in Figure 5. These queries consider both the instance document and its schema, which references the linkbases. Figure 4 shows a linkbase associated with the instance given in Figure 3. XPath+ defines an abbreviated axis syntax, similar to the one used by XPath. The Table 1 presents this syntax. The XPath+ grammar based on the EBNF language is presented in [7].

Table 1. Xpath+ Abbreviated Syntax

Abbreviated Syntax	Semantics
/***	selects the arcs among all the instance elements (e.g.: /inst.xml/***)
/<element>**	selects all the arcs of the element (e.g.: /inst/assests**)
///	is similar to the XPath "//" operator and its semantics are equivalent to the ones of the *link-source::* operator.
...	selects the nodes which the destination node is the context node links. It works similarly to the XPath ".." operator and its semantics are identical to the *link-destination::* operator .
[[x]]	selects the x-th element of a node list.

```
<?xml version="1.0"?>
<xbrl ... >
 <link:schemaRef xlink:type="simple" xlink:href="bcb_taxonomy.xsd"/>
 <bcb:assets id="id_assets" contextRef="c1" unitRef="u1">270190618</bcb:assets>
 <bcb:current_assets contextRef="c1" unitRef="u1">261376808</bcb:current_assets>
 <bcb:noncurrent_assets contextRef="c1"
      unitRef="u1">8813810</bcb:noncurrent_assets>
 <bcb:liabilities contextRef="c1" unitRef="u1">270190618</bcb:liabilities>
 <context id="c1"><!-- ... --></context>
 <unit id="u1"><!-- ... --></unit>
</xbrl>
```

Fig. 3. Instance document example – inst.xml

```
<link:linkbase . . .>
<definitionlink xlink:type="extended"
    xlink:role="http://www.xbrl.org/2003/role/link">
 <loc xlink:type="locator" xlink:href="bcb_taxonomy.xsd#assets"
     xlink:label="bcb_assets"/>
 <loc xlink:type="locator" xlink:href="bcb_taxonomy.xsd#current_assets"
     xlink:label="bcb_current_assets"/>
<definitionArc xlink:type="arc" xlink:show="replace" xlink:actuate="onRequest"
     xlink:from="bcb_assets" xlink:to="bcb_current_assets"
     xlink:arcrole="http://www.xbrl.org/2003/arcrole/general-special"/>
</definitionlink>
</linkbase>
```

Fig. 4. Linkbase example – definition.xml

```
Role: http://www.xbrl.org/2003/arcrole/general-special
Exist arc between current_assets and assets
Exist arc between noncurrent_assets and assets
```

Fig. 5. Example of link-destination use

2.4 *XPath+* Processor Architecture

Figure 6 presents the components of the XPath+ processor architecture. As input, the XPath+ processor receives one XML instance document, zero or more schemas that validate the instance documents, zero or more related linkbases and an XPath+ query expression. The main modules of the processor are: (1) the XPath+ parser processes the queries and is responsible for syntax checking as well; (2) the XML parser uses DOM [8] to create a memory representation of the input documents. It also validates these documents according to their schemas; (3) the expansion module analyses the XML documents in order to create their representation in memory using DOM. Then, it goes through them looking for links or linkbase references and inserts the link relations in the memory representation as well; (4) the optimization module is supposed to analyze the input queries and to rewrite them using optimization algorithms in order to reduce queries processing time. However this module has not been implemented yet; and (5) the executor module is the one that actually performs the query processing. It applies the input query to the documents memory model and returns the result as a node list, like XPath does. It is also able to interpret functions used in queries. The function library contains the list of all the predefined XPath functions, as well as the ones added by XPath+.

The XPath+ implementation is based on the Java platform. An extension of the org.w3c.dom.xpath package was created so that the processor can handle both XPath+

and XPath expressions. DOM was adopted because it is a W3C standard, and thus it is compatible with other XPath applications. The processor was implemented as a component, to enhance the possibility of system integration. XLink defines arc roles, in order to express the role played by the arcs when link traversing occurs. Two arcs may have the same label (defined by xlink:label) if they have different roles. The XPath+ processor deals with this issue. The complex nature of the link network may result in link circles. The XPath+ processor stores the visited linkbases and schemas. Thus it is able to detect when a second visit is done to the same document and it deals with it accordingly. An instance document may not contain all the elements defined on its schema, and some schema elements can be related and still not found in an instance document. XPath+ considers only the elements found in the instance.

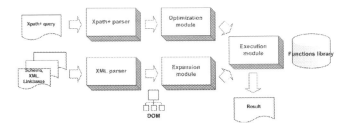

Fig. 6. The *XPath+* Processor Architecture

3   A Case Study Based on XBRL Documents

The need for XLink support by the XPath language is noticed in an example of an XBRL report, shown in Figure 3. This figure contains part of a balance sheet. Semantically, current assets and noncurrent assets are some of the items that form what accountants call assets. These three terms form a concept hierarchy. However, the XBRL instance document does not show this hierarchy, the linkbase plays this role. As Figure 4 shows, the linkbase uses arcs to define a generalization-specialization relationship, and thus a concept hierarchy is created.

This example highlights the need for link traversing. Using XPath, it is not possible to navigate over links, and thus it is not able, for exemple, to process a query regarding the children of the *bcb:assets* element. Consider that a company manager needs the value of the company assets stored in XBRL format. If he uses XPath to obtain this value, first he has to talk to the company's accountant. The accountant tells him that those elements that constitute the assets value are current and noncurrent assets. Then, he must open and analyse the XBRL file (see Figure 3) and look for elements, one by one. Once he is aware of the position of all the assets elements, he finally makes the query */xbrl[position() == 3] | /xbrl[position() == 4]*, which returns the third and fourth elements of the XBRL document. Using XPath, an expert must be involved in the querying process, to ensure that the correct elements are used. On the other hand, if that manager uses XPath+, he will only have to write the following query: */link-destination::bcb:assets*. Figure 6 shows the XPath+ query result.

As the concept hierarchy is in a separated XML file (the linkbase), XPath cannot identify that *bcb:current_assets* and *bcb:noncurrent_assets* are children of *bcb:assets*, and therefore an expert, the company's accountant, must provide the concept semantics. It is an additional work needed to handle XPath. The XPath+ processor is able to traverse arcs and get to the linkbase file, where the semantics are expressed. Therefore, XPath+ plays the expert's role in this context. Notice that in real applications the node tree is usually much larger, which makes XPath queries more complex. Using XPath+, link related query expressions are the same, whether executed on a long link intensive file or in a short link free file.

```
####################################################
#              XPathPlusJava Processor              #
####################################################

XPathPlusJava: c:\xlinktest\inst.xml/linkdestination::bcb:assets

Using file: c:\xlinktest\inst.xml
Using element: bcb:assets
Using namespace: www.bcb.gov.br
Using type: destinationArc
Instance Arcs:
Linkbases Arcs:
Role: http://www.xbrl.org/linkprops/linkRef/presentation
Exist arc between assets_current and assets
Exist arc between assets_nonCurrent and assets
Role: http://www.xbrl.org/linkprops/linkRef/calculation
Exist arc between assets_current and assets
Exist arc between assets_nonCurrent and assets
Role: http://www.xbrl.org/linkprops/linkRef/definition
Exist arc between assets_current and assets
```

Fig. 7. Screen result of the XPath+ query /linkdestination::bcb:assets

4 Related Work

Lizorkim and Lisovsky [9] defined several categories for XLink implementations. Our work fits in the category of Applications Programming Interfaces for link management. Lizorkim's and Lisovsky's solution concept is similar to our work. They use S-expressions in order to represent the XML document in a new format, SXML. S-expression is a native type in the functional programming language Scheme, which is one of the reasons for the adoption of this language. Scheme would avoid the impedance mismatch problem. However, link navigation engines are usually part of a larger architecture, written in an object-oriented language. Therefore, in our opinion, the use of Scheme only changes the impedance mismatch trigger location. Besides, the currently used object-oriented languages fully support XML trees. A example is Java and its DOM API. DOM is a W3C standard and thus can be reused in a number of applications. XPath+ is implemented based on Java and DOM.

May and Malheiro [10] use a logical data model of related XML documents in order to work with links. By using the namespace *dblink*, it is possible to specify the link behaviour when the query is made. The main difference between this approach and XPath+ is the possibility to handle links that point to distributed sources. However, in order to use this functionality, it is necessary to modify files, making this solution not practical to applications handling a large amount of XML data.

Laurent [11] proposes an approach using SAX [12], which creates a link collection that can take requests from the applications to know which elements contain links, as well the targets and behaviours of them. It is claimed that this way makes possible for users to gain convenience, as they will not need to build XLink processing into SAX handlers, thus helping the development of applications which deals with XLink.

XLink Processor [13] is a commercial solution based on Java and DOM and supports all kinds of XLink links. Another commercial solution based on Java and DOM to manipulate only XBRL links is the Batavia XBRL Java Library [14]. Java XBRL API Implementation [15] is an open source project, based on SAX, that provides an XLink processor for XBRL documents. SAX is more memory efficient than DOM, but it is not a W3C Recommendation. Besides to support any kind of relationship based on XML Schema and XLink, not just a subgroup as defined by XBRL, XPath+ is not a proprietary solution.

5   Conclusion and Future Work

In this paper we presented the XPath+, an XPath extension that supports XLink links traversing. Using XPath+, it is possible to navigate across link related documents. XPath does not support such navigation. The extension is especially useful to deal with linkbases, because it can associate linked XML elements. The XPath+ processor is based on Java and DOM. As both technologies are widely adopted, XPath+ can be used with plenty of existing applications. As an extension of XPath, the processor handles both types of expressions. We intend to optimize the navigation over links. One of the implementation problems we currently face is the execution time. We are also working on the specification of an access control module using XPath+.

References

1. XML Linking Language (XLink) Version 1.0, http://www.w3.org/TR/xlink
2. XML Path Language (XPath) 2.0, http://www.w3.org/TR/xpath20
3. Extensible Business Reporting Language (XBRL) 2.1, http://www.xbrl.org
4. XML Schema, http://www.w3.org/TR/2004/REC-xmlschema-0-20041028
5. Gottlob, G., Koch, C., Pichler, R.: XPath query evaluation: improving time and space efficiency. In: 19th International Conference on Data Engineering, pp. 379–390
6. Libkin, L.: Logics For Unranked Trees: An Overview. Logical Methods in Computer Science 2, 1–31 (2006)
7. Silva, P.C., Aquino, I.J.S., Times, V.C.: A Query Language For Navigation Over Links. In: XIV Simpósio Brasileiro de Sistemas Multimídia e Web (2008)
8. Document Object Model, http://www.w3.org/TR/2004/REC-DOM-Level-3-Core-20040407/
9. Lizorkim, D.A., Lisovsky, K.Yu.: The Query Language to XML Documents Connected by Link Links. Programming and Computer Software 31(3), 133–148 (2005)
10. May, W., Malheiro, D.: A Logical, Transparent Model for Querying Linked XML Documents (2003)
11. XLinkFilter, http://www.simonstl.com/projects/xlinkfilter/index.htm
12. Simple API for XML, http://www.saxproject.org/
13. XLiP, http://software.fujitsu.com/eninterstage-xwand/ activity/xbrltools/xlip/index.html
14. Batavia XBRL Java Library, http://www.batavia-xbrl.com
15. XBRLAPI Java XBRL API implementation, http://www.xbrlapi.org/

A Data Parallel Algorithm for XML DOM Parsing

Bhavik Shah[1], Praveen R. Rao[1], Bongki Moon[2], and Mohan Rajagopalan[3]

[1] University of Missouri-Kansas City
{BhavikShah,raopr}@umkc.edu
[2] University of Arizona
bkmoon@cs.arizona.edu
[3] Intel Research Labs
mohan.rajagopalan@intel.com

Abstract. The extensible markup language XML has become the de facto standard for information representation and interchange on the Internet. XML parsing is a core operation performed on an XML document for it to be accessed and manipulated. This operation is known to cause performance bottlenecks in applications and systems that process large volumes of XML data. We believe that parallelism is a natural way to boost performance. Leveraging multicore processors can offer a cost-effective solution, because future multicore processors will support hundreds of cores, and will offer a high degree of parallelism in hardware. We propose a *data parallel* algorithm called *ParDOM* for XML DOM parsing, that builds an in-memory tree structure for an XML document. *ParDOM* has two phases. In the first phase, an XML document is partitioned into chunks and parsed in parallel. In the second phase, partial DOM node tree structures created during the first phase, are linked together (in parallel) to build a complete DOM node tree. *ParDOM* offers fine-grained parallelism by adopting a flexible chunking scheme – each chunk can contain an arbitrary number of start and end XML tags that are not necessarily matched. *ParDOM* can be conveniently implemented using a data parallel programming model that supports map and sort operations. Through empirical evaluation, we show that *ParDOM* yields better scalability than PXP [23] – a recently proposed parallel DOM parsing algorithm – on commodity multicore processors. Furthermore, *ParDOM* can process a wide-variety of XML datasets with complex structures which PXP fails to parse.

1 Introduction

The extensible markup language XML has become the de facto standard for information representation and exchange on the Internet. Recent years have witnessed a multitude of applications and systems that use XML such as web services and service oriented architectures (SOAs) [16], grid computing, RSS feeds, ecommerce sites, and most recently the Office Open XML document standard (OOXML). Parsing is a core operation performed before an XML document can be navigated, queried, or manipulated. Though XML is simple to read and process by software, XML parsing is often reported to cause performance bottlenecks for real-world applications [22,32]. For example, in a SOA using web services technology, services are discovered, described, and invoked using XML messages [10]. These messages can reach up to several megabytes in size, and thus parsing can cause severe scalability problems.

Z. Bellahsène et al. (Eds.): XSym 2009, LNCS 5679, pp. 75–90, 2009.

Recently, high performance XML parsing has become a topic of considerable interest (*e.g.,* XML Screamer [19], schema-specific parser [11], PXP [23,24], Parabix [7]). XMLScreamer and schema-specific parser leverage schema information for optimizing tasks such as scanning, parsing, validation, and deserialization. On the other hand, PXP and Parabix exploit parallel hardware to achieve high XML parsing performance. Our work in this paper also exploits parallel hardware to achieve high parsing performance.

With the emergence of large-scale throughput oriented multicore processors [26][15], we believe parallelism is a natural way to boost the performance of XML parsing. Leveraging multicore processors can offer a cost-effective way to overcome the scalability problems, given that future multicore processors will support hundreds of cores, and thus, offer a high degree of parallelism in hardware. A data parallel programming model offer numerous benefits for future multicore processors such as expressive power, determinism, and portability [12]. For instance, traditional thread-based approaches suffer from non-deterministic behavior and make programming difficult and error prone. On the contrary, a program written in a data parallel language (*e.g.,* Ct [13]) has deterministic behavior whether running on one core or hundred cores. This eliminates data races and improves programmer productivity. Thus, there has been a surge of interest to develop data parallel models for forward scaling on future multicore processors [8,12].

With these factors in mind, we propose a *data parallel* XML parsing algorithm called *ParDOM*. In this paper, we focus on **XML DOM (Document Object Model) parsing** [30], because it is easy to use by a programmer and provides full navigation support to an application, and it is widely supported in open-source and commercial tools (*e.g.,* SAXON [18], Xerces [3], Intel Software Suite [1], MSXML [2]). Further, DOM parsing poses a fundamental challenge of parallel tree construction. Since DOM parsing requires documents to fit in main memory, we only consider XML documents that are of several megabytes in size.

ParDOM is a two-phase algorithm. In the first phase, an XML document is partitioned into chunks and are parsed in parallel. In the second phase, partial DOM node tree structures created during the first phase, are linked together (in parallel) to build a complete DOM node tree in memory. Our algorithm offers fine-grained parallelism by adopting a flexible chunking scheme. Unlike a previous parallel algorithm called PXP [23,24], wherein chunks represent subtrees of a DOM tree, *ParDOM* creates chunks that can contain an arbitrary number of start and end XML tags that are not necessarily matched. *ParDOM* can be conveniently implemented using a data parallel programming model that supports `map` and `sort` operators. Through empirical evaluation, we show that *ParDOM* yields better scalability than PXP on commodity multicore processors. Furthermore, *ParDOM* can process a wide-variety of XML datasets with complex structures which PXP fails to parse.

2   Background and Motivation

2.1   XML Documents and Parsing Techniques

An XML document contains elements that are represented by start and end element tags. Each element can contain other elements and values. An element can have a list of (attribute, value) pairs associated with it. An XML document can be modeled as an ordered labeled tree. A well-formed XML document follows the XML syntax rules. For example, each element has a start tag and a matching end tag.

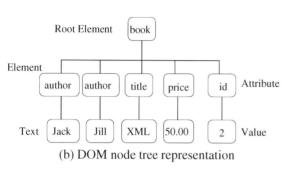

(b) DOM node tree representation

Fig. 1. Example

For an XML document to be accessed and manipulated, it should first be parsed. Many XML parsing models have been developed that trade off between the ease of use, APIs exposed to applications, memory footprint, parsing speed, and support for XPath [5].

Among these, DOM parsing and SAX parsing are widely supported. Document Object Model (DOM) [30] parsing builds an in-memory tree representation of an XML document by storing its elements, attributes, and values along with their relationships. (Other DOM node types have been defined by W3C [30]. We restrict ourselves to the most common ones: Element, Attribute, Text/Value.) A DOM node tree aids easy navigation of XML documents and supports XPath [5]. A DOM tree for a document is shown in Figure 1. The order of siblings in the tree follows the order in which their elements appear in the document (a.k.a. document order).

SAX parsing [21] is an event based parsing approach. It is light-weight, fast, and requires a smaller memory footprint than DOM parsing. However, an application is responsible for maintaining an internal representation of a document if required. Newer parsing models such as StAX [6] and VTD-XML were developed to improve over DOM and SAX. The Binary XML standard [14], though not a parsing model, was proposed to reduce the verbosity of XML documents and the cost of parsing. However, human-readability is lost.

2.2 Prior Work on Parallel XML Parsing

Recently, Pan *et al.* proposed a parallel XML DOM parsing algorithm called PXP for multicore processors [23]. This approach first constructs a skeleton of a document in memory. Using the skeleton, the algorithm identifies chunks of the document that can be parsed in parallel. (Each chunk denotes a subtree of the final DOM tree.) This task requires recursively traversing the skeleton until enough chunks are created. After the chunks are created, they are parsed in parallel to create the DOM tree. Subsequently, Pan *et al.* proposed an improved algorithm to parallelize the skeleton construction [24].

However, these algorithms have the following shortcomings that motivate our research. First, the skeleton requires extra memory that is proportional to the number of node in the DOM tree. Further, the partitioning scheme based on subtrees can cause load imbalance on processing cores for XML documents with irregular or deep tree structures (*e.g.*, TREEBANK with parts-of-speech tagging [29]). This scheme severely limits the granularity of parallelism that can be achieved, and thus cannot scale with increasing core count.

Wu *et al.* proposed a parallel approach XML parsing and schema validation [31]. Although their chunking scheme during parsing is similar to that of *ParDOM*, the partial DOM trees for each chunk are linked sequentially during post-processing. Parabix [7], though not a parallel DOM parsing algorithm, exploits parallel hardware for speeding up parsing by scanning the document faster. Rather than reading a byte-at-a-time from an XML document, Parabix fetches and processes many bytes in parallel.

2.3   Prior Work on Data Parallel Programming Models

The emergence of multicore processors demands new solutions for expressing parallelism in software to fully exploit their capabilities [4]. There has been a keen interest in developing parallel programming models for this purpose. Intel's Ct [13] supports a data parallel programming model and aims on forward scaling for future multicore processors. Data Parallel Haskell is another effort to exploit the power of multicores [8].

In recent years, programming models to support large-scale distributed computing on commodity machines have been developed. The MapReduce paradigm and associated implementation was introduced by Google for performing data intensive computations that can be distributed across thousands of machines [9]. Hadoop (`http://hadoop.apache.org`) and Disco (`http://discoproject.org`) are two different open source implementations of MapReduce. Phoenix [25] is a shared memory MapReduce implementation. Recently, a distributed execution engine called Dyrad [17] was proposed for coarse-grained data parallel applications.

3   Our Proposed Approach

We begin with a description of a serial algorithm for building a DOM tree. We present a scenario to motivate the design of our parallel algorithm *ParDOM*. We focus on XML documents whose DOM trees can fit in main memory. (For very large XML documents, other parsing models (*e.g.*, SAX [21]) should be used.) For ease of exposition, we focus on elements, attributes, and text/values in XML documents. Although a text can appear anywhere within the start and end tag of an element, we shall first assume that it is strictly enclosed by start and end element tags, *e.g.*, `<author>Jack</author>`. Later in Section 4.4, we will discuss how to handle the case `<author>US<first>Jack</first>English</author>`. Here `US` and `English` are text associated with `author` according to the XML syntax.

3.1   A Serial Approach

A DOM tree can be built by extracting tokens (*e.g.*, start and end tags) from a document by reading it from the beginning. A stack S is maintained and is initially empty. This stack essentially stores the information of all the ancestors (in the DOM tree) of the current element being processed in the document. When a start element tag say `<e>` is read, a DOM node d_e is created for element `e` and any (attribute,value) pair that is associated with the element is parsed and stored, by creating the necessary DOM nodes. If S is not empty, then this implies that d_e's parent node has already been created. Node d_e is linked as the rightmost child of its parent by consulting the top of stack S.

(The order of siblings follows the order in which the elements appear in the document.) The pair (d_e, e) is pushed onto the stack S. If e encloses text, then a DOM node for the text is also created and linked as a "text" child of d_e. When an end element tag say $</\text{e}>$ is read, e is checked with the top of stack S. If the element names do not match, then the parsing is aborted as the document is not well-formed. Otherwise, the top of S is popped and the parsing continues. After the last character of the document is processed, if S is empty, then the entire DOM tree has been constructed. Otherwise, the document is not well-formed.

3.2 A Parallel Approach

Given an XML document, any data parallel algorithm would perform the following tasks: (a) construct partial DOM structures on chunks of the XML document, and (b) link the partial DOM structures. Suppose n processor cores are available, each core can be assigned a set of chunks. Each core then processes one chunk at-a-time and establish parent-child links as needed.

Example 1. Figure 2 shows three chunks 0, 8, and 20 whose partial DOM trees have been constructed. Suppose elements Y and Z are child elements of X. The parent-child links between them have been created as shown.

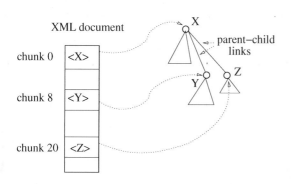

Fig. 2. Partial DOM construction & linking process

Motivating Scenario: If the linking tasks were to be done concurrently with the partial DOM construction tasks, then *synchronization* is necessary to ensure that parent-child links are updated correctly without race conditions. (Note that according the XML standard, there is an ordering among siblings based on their relative positions in the input document.) It is also possible that a parent DOM node has not been created yet, while its child DOM node (present in a subsequent chunk) has already been created. As a result, an attempt to create a link to the parent would have to wait. Mutexes can be used for the purpose of synchronization. But can synchronization primitives be avoided altogether? We believe this is possible, if we design a two-phase parallel algorithm. In the first phase, partial DOM structures are created in parallel over all the chunks. Once all the chunks have been processed, in the second phase, for each parent node, with at least one child in a different chunk, all its child nodes appearing in subsequent chunks are grouped together. Each group is processed by a single task, and all the missing parent-child links are created. Such tasks can be executed in parallel.

Challenges in ParDOM: Two challenges arise in the design of our two-phase parallel algorithm. First, to obtain fine-grained parallelism, each chunk should be created using a criteria independent of the underlying tree structure of a document. Second, the partial DOM structure (created for a chunk) must be located and linked correctly in the final DOM tree.

To address the first challenge, *ParDOM* adopts a flexible chunking scheme – each chunk contains an arbitrary number of start and end tags that are not necessarily matched. The required chunk size can be specified in many ways such as (a) the number of bytes per chunk, (b) the number of XML tags per chunk, or (c) the number of start tags per chunk. (We ensure that a start tag, end tag, or text is not split across different chunks.)

Example 2. Consider an XML document in Figure 3. It is partitioned into three chunks where the ith chunk $(i \geq 0)$ starts from the $(3 * i + 1)^{th}$ start element tag.

To address the second challenge, *ParDOM* uses a simple numbering scheme for XML elements and a stack P that stores the element numbers and names. Numbering schemes were proposed in the past for indexing and querying XML data (*e.g.,* Extended-preorder [20], Dewey [27]). Essentially, each element is assigned a unique id. Relationships between elements (*e.g.,* parent-child, ancestor-descendant, sibling) in an XML document tree can be inferred from their ids. *ParDOM* uses *preorder numbering*, where each element's id is the preorder number of its node in the XML document tree. The ids can be computed on-the-fly while extracting tokens from a document. Starting with a counter value of 0, each time a start element tag is seen, the counter is incremented and its value is the preorder number of the element. The root element is thus assigned the preorder number 1. In Figure 3, elements `book`, `last`, and `title` are assigned preorder numbers 1, 4, and 7, respectively.

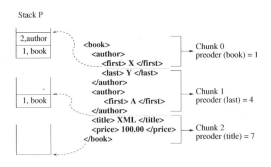

Fig. 3. Three chunks and the state of stack P

While preorder numbers can be used to determine the ordering among siblings (by sorting their ids), they cannot determine parent-child or ancestor-descendant relationships between elements. The parent-child relationship between elements is inferred using the stack P that is maintained similarly to stack S described in Section 3.1. Each entry in P is a pair *(id,element)*. Suppose the serial algorithm is applied to an input document. When a new chunk is read, the top of stack P, if P is not empty, denotes the element in some previous chunk whose end tag has not yet been encountered. In addition, exactly one entry in P denotes the parent of the first start element tag that appears in the current chunk (except for chunk 0).

Example 3. In Figure 3, the ids of the first elements in each chunk are shown. After chunk 0 is processed, the state of stack P is shown. The top element `author` in P

denotes the parent of `last` that appears in chunk 1. Similarly, the state of P is shown after processing chunks 1 and 2.

When a chunk is parsed independently, if the state of stack P is known just after processing the previous chunk, then the parent of every element in the chunk can be determined. Thus the partial DOM structure constructed for the chunk can be correctly linked to the final DOM tree. At first glance, it may seem that each chunk should be parsed serially for correctness. However, this is not the case – only stack P should be correctly initialized, and this can be done without actually constructing partial DOM trees for a chunk.

One approach is to first read the entire document, compute preorder numbers (or ids) of elements and update the stack P appropriately. At each chunk boundary, the stack P is copied and stored. We call this copy of P a *chunk boundary stack*. Once all *chunk boundary stacks* are created, the chunks can be parsed in parallel. Note that to link the partial DOM structures into the final DOM tree, the references to DOM nodes of elements whose end tags were not present in the chunk should be maintained.

4   Implementing *ParDOM*

ParDOM can be conveniently implemented in a data parallel programming model that supports map and sort operators. Given a sequence of items, a map operation applies a function f to each item in the sequence. Parallelism can be exploited for both the map and sort operators. For subsequent discussions, we will use the term "*a map task*" to refer to a map operator being applied to a single item in a sequence.

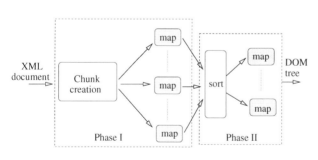

Figure 4 shows the overall sequence of tasks performed by *ParDOM*. Phase I begins with chunk creation that includes establishing chunk boundaries, assigning preorder numbers to elements, and creating chunk boundary stacks. Then the map tasks are run in parallel on all the chunks – each map task constructs par-

Fig. 4. Sequence of tasks in *ParDOM*

tial DOM trees on its chunk. Note that as soon as the boundary of a chunk is established and its chunk boundary stack is constructed, a map task can be executed on that chunk. A map task also outputs information regarding those elements whose parents appear in some preceding chunks along with their parent ids. Once all the map tasks complete, in Phase II, the information output by the map tasks are grouped according to the parent node ids, using a sort operation. For each parent id, its group is processed by exactly one map task. A map task creates missing parent-child links between a parent DOM node and all its child DOM nodes in the group. It also ensures that siblings are in document

---

Algorithm 1. Chunk creation

Global: int $nodeId \leftarrow 0$; int $chunkId \leftarrow 0$; intArray[] $firstNodeId$; stack P; stackArray[] P_c;

    procedure ChunkCreate($dataIn$, size)
1: $begin \leftarrow dataIn$;
2: $end \leftarrow begin + size + \delta$; /* avoid splitting XML tags and going beyond EOF */
3: **foreach** $(e, type) \in [begin, end]$ **do**
4:      **switch** *type* **do**
5:          **case** *START:*
6:              **if** *first START tag in chunk* **then**
7:                  $P_c[chunkId] \leftarrow P$; /* Copy stack P */
8:                  $nodeId$++; /* Next preorder number */
9:                  $firstNodeId[chunkId] \leftarrow nodeId$;
              end
10:            $P.push(nodeId, e)$;
11:            **break**;
12:          **case** *END:* $P.pop()$; **break**;
13:          **otherwise** do nothing;
         end
      end
14: $chunkId$++;
15: $dataIn \leftarrow end + 1$;

---

order. Since *exactly one* map task creates the missing parent-child links, no locks are needed. Next, we describe the algorithmic details of each phase in *ParDOM*.

4.1 Phase I - Chunk Creation

The steps performed during chunk creation are shown in Algorithm 1. Each invocation of $ChunkCreate()$ identifies the boundaries of a single chunk, computes preorder numbers for the elements in it, and constructs its chunk boundary stack. The global variables are used for preorder numbering of elements and for storing chunk boundary stacks. The input arguments are $dataIn$, that points to the beginning of the current chunk, and a suggested chunk size. Lines 1-2 set up the chunk boundaries, where δ is chosen to ensure that a start tag, end tag, or text is not split across two chunks, and that the last chunk does not span beyond the end-of-file. Line 3 simply denotes tokenization of the chunk based on start and end XML tags. (The attributes and text/values are not needed at this stage and are ignored.) As the document is processed, stack P is copied and stored when the first start tag is encountered in a chunk (Line 7). Thus, a chunk boundary stack $P_c[chunkId]$ is created. (This differs slightly from our earlier discussion where P would have been copied at the beginning of a chunk.) In addition, the preorder number assigned to this element is stored (Line 9) so that during the execution of map tasks in Phase I, the element ids can be regenerated correctly. Finally, on Line 15, $dataIn$ is initialized to the beginning of the next chunk. The next invocation of $ChunkCreate()$ uses $dataIn$ as its input. Whether a document is well-formed or not can be checked during chunk creation.

4.2   Phase I - Partial DOM Construction

Once Algorithm 1 completes on a chunk, a map task processes that chunk to create partial DOM trees. Algorithm 2 describes the steps involved. A local stack T, initially empty, is used to store an element's id and a reference to its DOM node. It is updated similar to stack P.

When a start of an element e is encountered, a DOM node is created, and the (attribute,value) pairs are processed and stored (Line 6). If T is empty, then e's parent is in some previous chunk. The parent of e is known from the top entry of the chunk boundary stack. A key-value pair is output where the key denotes the parent of e and the value is a reference to the DOM node for e (Lines 9-10). If T is not empty, then e's parent is the top entry of T. The DOM node for e is added as the rightmost child of its parent (Line 11).

When an end of an element e is encountered, stack T is checked. If T is empty, then e's start tag was present in some previous chunk. (Note that T cannot be empty at this point for chunk 0 if the document is well-formed.) The chunk boundary stack is updated if a start tag was already encountered while processing this chunk (Line 17). When a text is encountered, it is associated with its element using stack T (Line 21).

Finally, we pop all entries in T (Lines 24-28). These correspond to elements whose end tags were absent in the current chunk, and thus may have child elements in subsequent chunks. To link an element's DOM node correctly to a child node, a reference to it should be available in Phase II. To achieve this, a key-value pair is output where the key is the element's id and value is a special DOM node that contains the reference to its actual DOM node (Line 27). This is done to distinguish this special node from other DOM node references corresponding to child nodes output in Line 10.

Example 4.   The partial DOM tree structures are shown in Figure 5 for the chunks in Figure 3. The key-value pairs are output for chunk 1 and chunk 2. The key-value pairs output in Line 27 are not shown.

4.3   Phase II - Linking Partial DOM Trees

The linking process is straightforward. The key-value pairs output in Phase I are sorted by the key *i.e.,* parent id. (The value component denotes a reference to a DOM node.) For each group of key-value pairs with the same key, a map task creates parent-child links between DOM nodes,

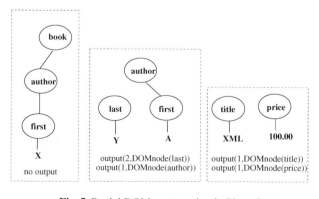

Fig. 5. Partial DOM construction in Phase I

Algorithm 2. Map task for Phase I in *ParDOM*

---

   procedure MapPhaseI($begin, end, chunkId$)
1: stack T; /* Each entry contains a DOM node ptr and node id */
2: $nodeId \leftarrow firstNodeId[chunkId]$;
3: **foreach** $(e, type) \in [begin, end]$ **do**
4:     **switch** *type* **do**
5:        **case** *START:*
6:           create DOM node for element e including its attributes, and also store
              $nodeId$
7:           let d_e denote a reference to e's DOM node
8:           **if** T *is empty* **then**
9:              $(parentId, tag) \leftarrow P_c[chunkId].top()$
10:             $Output(parentId, d_e)$ /* Like emitIntermediate() of MapReduce */
           else
11:             add d_e as the right most child of DOM node referenced by $T.top()$
           end
12:           $T.push(d_e, nodeId)$;
13:           $nodeId$++; **break**;
14:        **case** *END:*
15:           **if** T *is EMPTY* **then**
16:             **if** *a START tag was seen in chunk* **then**
17:                $P_c[chunkId].pop()$;
             end
           else
18:             $T.pop()$;
           end
19:           **break**;
20:        **case** *TEXT:*
21:           store text as child of DOM node referenced by $T.top()$;
22:           **break**;
23:        **otherwise** do nothing;
     end
24:     **while** T *is* $\overline{EMPTY}$ **do**
25:        $(nodeId, d_e) \leftarrow T.top()$
26:        create a special node d_* containing the reference d_e
27:        $Output(nodeId, d_*)$ /* Like emitIntermediate() of MapReduce */
28:        $T.pop()$
     end
  end

---

and ensures that the child DOM nodes are in document order. Each DOM node stores its node id and can be ordered by sorting on the node id. In the interest of space, the algorithm is not outlined here.

Example 5. The partial DOM structures in Figure 5 are linked during phase II. The DOM nodes for author, title, and price are linked as child nodes of book (with id 1) after sorting them based on their node ids. The DOM node for last is linked to author (with id 2).

4.4 Extensions and Memory Requirement

To support text that are not strictly enclosed within a start and end tag the following modifications are needed. If the element containing the text appears in the same chunk, then it is linked to the text node. Otherwise, Algorithm 2 should be modified to output a key-value pair (similar to Line 10) when a text appears as the first item. The parent is known from the chunk boundary stack. In Phase II, this text will be linked to its element DOM node.

In *ParDOM*, the additional memory required to store chunk boundary stacks depends on the number of chunks and the maximum depth of the document tree. On the contrary, PXP [23] consumes additional memory that is linear in the number of tree nodes for skeleton construction.

5   Experimental Results

We compared *ParDOM* with PXP [23] – a data parallel DOM parsing algorithm. We obtained a Linux binary for PXP from the authors. All experiments were conducted on a machine running Fedora 8 with a Intel Core 2 Quad processor (2.40GHz). The machine had 2GB RAM and 500GB disk space.

5.1   Using MapReduce to Implement *ParDOM*

We implemented *ParDOM* using Phoenix [25], which is a shared memory MapReduce implementation written in C. The code was compiled using the GNU gcc compiler version 4.0.2. The MapReduce model provides a convenient way for expressing the two phases of *ParDOM*. This model has two phases, namely, the Map phase and the Reduce phase. The input data is split, and each partition is provided to a Map task. Each Map task can generate a set of key-value pairs. The intermediate key-value pairs are merged and automatically grouped based on their key. In the Reduce phase, each intermediate key along with all the associated values is processed by a Reduce task. A MapReduce program written in Phoenix allows a user-defined split(), map(), and reduce() procedures. In our MapReduce implementation of *ParDOM*, split() implemented Algorithm 1, map() implemented Algorithm 2, and reduce() implemented the steps described in Section 4.3.

5.2   *ParDOM* vs. PXP

ParDOM and PXP were evaluated on a variety of XML datasets with different structural characteristics and sizes.[1] These datasets were obtained from University of Washington [29]. Figure 6 shows the characteristics of each dataset in terms of its size, number of elements and attributes, and maximum tree depth. DBLP contains computer science bibliographic information. SWISSPROT is a curated protein sequence database. TREEBANK captures linguistic structure of a Wall Street Journal article using parts-of-speech tagging. It has deep, irregular structure. LINEITEM contains data from the TPC-H Benchmark [28].

PXP requires scanning the input document during a preparsing phase for constructing a skeleton of the document. A skeleton is a light-weight representation of the document's structure and does not involve the creation of DOM tree nodes. Then the document is partitioned into tasks (denoted by subtrees) using the skeleton, and these

---

[1] These datasets are different from those used by the authors of PXP [23].

tasks are run in parallel to create partial DOM trees. Preparsing and task partitioning are performed sequentially. Finally, PXP requires a postprocessing phase to remove some temporary DOM nodes.

ParDOM also requires scanning the input document during chunk creation (Algorithm 1). However, a careful implementation in Phoenix allows us to interleave the chunk creation phase with the Map tasks in Phase I. Note that once a chunk boundary stack is computed for a chunk, it is ready to be processed by a Map task.

Dataset	Size	Max depth	# of elements	# of attributes
DBLP	127MB	6	3332130	404276
SWISSPROT	109MB	5	2977031	2189859
TREEBANK	82MB	36	2437666	1
LINEITEM	30MB	3	1022976	1

Fig. 6. XML datasets and their characteristics

Measurements & Results. For each dataset, we ran *ParDOM* and PXP on 2, 3, and 4 cores. For *ParDOM*, chunks were created by specifying *bytes per chunk*, and each chunk was extended to contain the nearest end tag of an element. The PXP code provided to us could not process XML documents beyond a certain size and crashed during preparsing. Therefore, we created smaller datasets of size 8MB, 16MB, and 32MB from our original datasets. We measured the wall-clock time and computed the average over three runs. Each dataset was read once before parsing so that it is cached in the file system buffer to avoid I/O while parsing.

To compute speedup, we ran a serial parsing algorithm (Section 3.1) and PXP on one core. Let us call them as T_s and T_{PXP}, respectively. *ParDOM*'s speedup was measured by computing the ratio of T_s with its parallel parsing time. (The parallel parsing time included the cost of chunk creation.) PXP's speedup was measured by computing the ratio of T_{PXP} with its parallel parsing time. (The parallel parsing time included the cost of preparsing.)

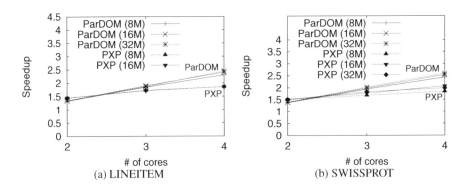

Fig. 7. Speedup measurements

Speedup: Figure 7(a) and 7(b) show the speedup of *ParDOM* and PXP for LINEITEM and SWISSPROT, respectively. The chunk size of 256KB was selected for *ParDOM*, beyond which the parallel parsing time did not improve significantly. Clearly, *ParDOM* had better speedup than PXP at 4 cores for both LINEITEM and SWISSPROT. *ParDOM* achieved a speedup of around 2.5 with 4 processing cores. (Note that PXP crashed for 32MB of LINEITEM dataset during preparsing phase, and hence is not shown in the plot.) Interestingly, PXP failed to parse TREEBANK and DBLP even for 8MB dataset sizes and crashed. The crash occurred in the preparsing phase. In these datasets, the fanout at nodes other than the root were not large. Further, TREEBANK had deep tree structures. This clearly demonstrates the superiority of *ParDOM* over PXP for parallel DOM parsing as it can process a variety of tree structures and document sizes.

Figure 8(a) shows the speedup for *ParDOM* on all the four datasets, each of size 64MB. We achieved the best speedup of 2.61. We observed similar trends in the speedup for *ParDOM* when the original datasets in Figure 6 were used.

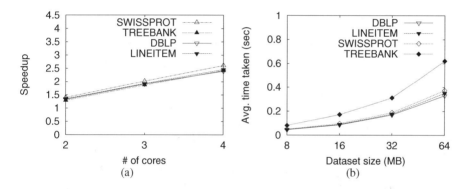

Fig. 8. (a) Speedup of *ParDOM* (64MB). (b) Data scalability.

Data Scalability: To measure how *ParDOM* scales with increase in dataset size, we measured the average parsing time (over 3 runs) for datasets size of 8MB, 16MB, 32MB, and 64MB. The results for 4 cores is plotted in Figure 8(b). For instance, *Par-DOM* required 0.312 secs and 0.621 secs to process 32MB and 64MB of TREEBANK, respectively.

For *ParDOM*, we measured the effectiveness of our simple chunking scheme on the distribution of load among the Map tasks in Phase I. We used the original datasets in Figure 6. A Map task that processed more elements created more DOM nodes. Figure 9 shows the mean and standard deviation of the number of elements processed per Map task excluding the last Map task that can have a smaller chunk size. We observed that for TREEBANK and LINEITEM the load was well-balanced among Map tasks as compared to DBLP and SWISSPROT. This is evident from the smaller σ values. DBLP and SWISSPROT datasets contained text of varied lengths that resulted in higher σ values. Thus chunking based solely on *bytes per chunk* may not be ideal for such datasets.

We also measured the load during Phase II of *ParDOM*, by considering the number of child nodes that were linked per task, excluding the root node. (The root node of each dataset had very large fanout.) The total, mean, and standard deviation for the number of child links created are shown in Figure 9. Note that more tasks were required for TREEBANK as compared to the other datasets because an average of 1.5 child nodes were linked per task. SWISSPROT had larger fanout among nodes as compared to DBLP and this is reflected in the total number of child nodes that were linked in Phase II.

Dataset	# of elements per Map task Phase I		Total # of parent–child links created in Phase II		
	Mean	σ	Total	Mean	σ
DBLP	24338.6	1545.3	1670	5.3	5.6
SWISSPROT	22789.8	725.0	4155	9.2	16.4
TREEBANK	23323.3	274.6	1622	1.5	0.9
LINEITEM	29041.9	33.1	425	5.4	4.8

Fig. 9. Load measurement

Load Balancing: Finally, we measured how much time was spent in the Map and Reduce phases in our *ParDOM* implementation. We used the original datasets for this experiment. We observed that in all cases the Reduce phase consumed less than 8% of the total time.

6   Conclusions

ParDOM is a data parallel XML DOM parsing algorithm that can leverage multicore processors for high performance XML parsing. *ParDOM* offers fine-grained parallelism by using a flexible chunking scheme that is oblivious to the structure of the XML document. *ParDOM* can be conveniently implemented in a data parallel language that supports `map` and `sort` operations. Our empirical results show that *ParDOM* provides better scalability than PXP [23] on commodity multicore processors. Further, it can process a wide variety of datasets as compared to PXP.

Acknowledgments. We thank the authors of PXP for their code and the anonymous reviewers for their insightful comments.

References

1. Intel XML Software Suite Performance Paper,
   http://intel.com/software/xmlsoftwaresuite
2. Microsoft XML Core Services (MSXML),
   http://msdn.microsoft.com/en-us/xml/
3. Xerces-C++ XML Parser, http://xerces.apache.org/xerces-c/
4. Asanovic, K., Bodik, R., Catanzaro, B.C., Gebis, J.J., Husbands, P., Keutzer, K., Patterson, D.A., Plishker, W.L., Shalf, J., Williams, S.W., Yelick, K.A.: The landscape of parallel computing research: A view from berkeley. Technical Report UCB/EECS-2006-183, EECS Department, University of California, Berkeley (December 2006)
5. Berglund, A., Boag, S., Chamberlin, D., Fernandez, M.F., Kay, M., Robie, J., Simon, J.: XML path language (XPath) 2.0 W3C working draft 16. Technical Report WD-xpath20-20020816, World Wide Web Consortium (August 2002)

6. Cable, L., Chow, T.: JSR 173: Streaming API for XML (2007), http://jcp.org/en/jsr/detail?id=173
7. Cameron, R.D., Herdy, K.S., Lin, D.: High performance XML parsing using parallel bit stream technology. In: CASCON 2008: Proc. of the 2008 conference of the center for advanced studies on collaborative research, New York, pp. 222–235 (2008)
8. Chakravarty, M.M.T., Leshchinskiy, R., Jones, S.P., Keller, G., Marlow, S.: Data Parallel Haskell: a status report. In: Proc. of the 2007 Workshop on Declarative Aspects of Multicore Programming, Nice, France, January 2007, pp. 10–18 (2007)
9. Dean, J., Ghemawat, S.: MapReduce: Simplified Data Processing on Large Clusters. In: Proc. of the OSDI 2004, San Francisco, CA (December 2004)
10. Engelen, R.A.V.: A framework for service-oriented computing with C and C++ Web service components. ACM Transactions on Internet Technology 8(3), 1–25 (2008)
11. Gao, Z., Pan, Y., Zhang, Y., Chiu, K.: A high performance schema-specific xml parser. In: IEEE Intl. Conf. on e-Science and Grid Computing, December 2007, pp. 245–252 (2007)
12. Ghuloum, A., Smith, T., Wu, G., Zhou, X., Fang, J., Guo, P., So, B., Rajagopalan, M., Chen, Y., Chen, B.: Future-proof data parallel algorithms and software on intel multi-core architecture. Intel Technology Journal 11(4), 333–348 (2007)
13. Ghuloum, A., Sprangle, E., Fang, J., Wu, G., Zhou, X.: Ct: A Flexible Parallel Programming Model for Tera-scale Architectures, 2007. Intel White Paper (2007)
14. Goldman, O., Lenkov, D.: XML Binary Characterization. Technical report, World Wide Web Consortium (March 2005)
15. Grohoski, G.: Niagara 2: A highly threaded server-on-a-chip. In: 18th Hot Chips Symposium (August 2006)
16. Huhns, M., Singh, M.P.: Service-Oriented Computing: Key Concepts and Principles. IEEE Internet Computing 9(1), 75–81 (2005)
17. Isard, M., Budiu, M., Yu, Y., Birrell, A., Fetterly, D.: Dryad: distributed data-parallel programs from sequential building blocks. In: Proc. of the 2nd ACM SIGOPS/EuroSys European Conference on Computer Systems 2007, pp. 59–72 (2007)
18. Kay, M.: SAXON: The XSLT and XQuery Processor, http://saxon.sourceforge.net
19. Kostoulas, M.G., Matsa, M., Mendelsohn, N., Perkins, E., Heifets, A., Mercaldi, M.: XML screamer: an integrated approach to high performance XML parsing, validation and deserialization. In: Proc. of the 15th International Conference on World Wide Web, New York, pp. 93–102 (2006)
20. Li, Q., Moon, B.: Indexing and querying XML data for regular path expressions. In: Proc. of the 27th VLDB Conference, Rome, Italy, September 2001, pp. 361–370 (2001)
21. Megginson, D.: Simple API for XML, http://sax.sourceforge.net/
22. Nicola, M., John, J.: XML parsing: a threat to database performance. In: Proc. of the 12th International Conference on Information and Knowledge Management, pp. 175–178 (2003)
23. Pan, Y., Lu, W., Zhang, Y., Chiu, K.: A Static Load-Balancing Scheme for Parallel XML Parsing on Multicore CPUs. In: Proc. of the 7th International Symposium on Cluster Computing and the Grid (CCGRID), Washington D.C., May 2007, pp. 351–362 (2007)
24. Pan, Y., Zhang, Y., Chiu, K.: Simultaneous transducers for data-parallel XML parsing. In: Proc. of Intl. Symposium on Parallel and Distributed Processing, April 2008, pp. 1–12 (2008)
25. Ranger, C., Raghuraman, R., Penmetsa, A., Bradski, G., Kozyrakis, C.: Evaluating MapReduce for Multi-core and Multiprocessor Systems. In: Proceedings of the 13th International Symposium on High-Performance Computer Architecture (HPCA), Phoenix, AZ (Feburary 2007)
26. Seiler, L., Carmean, D., Sprangle, E., Forsyth, T., Abrash, M., Dubey, P., Junkins, S., Lake, A., Sugerman, J., Cavin, R., Espasa, R., Grochowski, E., Juan, T., Hanrahan, P.: Larrabee: a many-core x86 architecture for visual computing. ACM Trans. Graph. 27(3), 1–15 (2008)

27. Tatarinov, I., Viglas, S.D., Beyer, K., Shanmugasundaram, J., Shekita, E., Zhang, C.: Storing and Querying Ordered XML Using a Relational Database System. In: Proc. of the 2002 ACM-SIGMOD Conference, June 2002, pp. 204–215 (2002)
28. TPC. TPC-H (2002), http://www.tpc.org/tpch/
29. UW XML Repository (2001),
    http://www.cs.washington.edu/research/xmldatasets
30. W3C. The document object model (1998), http://www.w3.org/DOM
31. Wu, Y., Zhang, Q., Yu, Z., Li, J.: A Hybrid Parallel Processing for XML Parsing and Schema Validation. In: Proceedings of Balisage Markup Conference (2008)
32. Zhang, J., Lovette, K.: XimpleWare W3C Position Paper. In: W3C Workshop on Binary Interchange of XML Information Item Sets (2003)

Optimizing XML Compression

Gregory Leighton and Denilson Barbosa

University of Alberta
Edmonton, AB, Canada
{gleighto,denilson}@cs.ualberta.ca

Abstract. The eXtensible Markup Language (XML) provides a powerful and flexible means of encoding and exchanging data. As it turns out, its main advantage as an encoding format (namely, its requirement that all open and close markup tags are present and properly balanced) yields also one of its main disadvantages: verbosity. XML-conscious compression techniques seek to overcome this drawback. Many of these techniques first separate XML structure from the document content, and then compress each independently. Further compression gains can be realized by identifying and compressing together document content that is highly similar, thereby amortizing the storage costs of auxiliary information required by the chosen compression algorithm. Additionally, the proper choice of compression algorithm is an important factor not only for the achievable compression gain, but also for access performance. Hence, choosing a compression configuration that optimizes compression gain requires one to determine (1) a partitioning strategy for document content, and (2) the best available compression algorithm to apply to each set within this partition. In this paper, we show that finding an optimal compression configuration with respect to compression gain is an **NP**-hard optimization problem. This problem remains intractable even if one considers a single compression algorithm for all content. We also describe an approximation algorithm for selecting a partitioning strategy for document content based on the branch-and-bound paradigm.

1 Introduction

The *eXtensible Markup Language (XML)* has become increasingly popular as a data encoding format. XML has many benefits, but one notable weakness: its verbosity, resulting from the high markup-to-content ratio imposed in large part by requiring every markup tag to be properly closed. The increasing size of XML datasets has motivated researchers to seek ways to reduce storage costs by applying compression techniques. Because XML is inherently a textual format, the naive solution is to apply a generic text compression scheme. However, such schemes are not aware of XML syntax, and therefore cannot easily exploit redundancies in the tree structure unambiguously induced by the proper nesting of markup tags inside the XML document (such as repeated subtrees), or even distinguish an element tag from a text segment. Thus, such a strategy severely hinders query processing, which is fundamentally based on traversing the structure of the document.

Z. Bellahsène et al. (Eds.): XSym 2009, LNCS 5679, pp. 91–105, 2009.

With such shortcomings in mind, many *XML-conscious* compression techniques have been proposed in recent years. Among them, *homomorphic approaches* to XML compression (e.g., [1–6]) preserve the original tree structure in the compressed representation by processing each node as it occurs during a pre-order traversal. *Permutation-based approaches* (e.g., [7–11]) re-arrange the document before performing compression, in an attempt to group "similar" nodes together and therefore improve the achievable compression rate. A commonly used permutation strategy treats structure separately from content, and then applies a partitioning strategy to group content nodes into a series of *data containers*. However, there is an inherent tradeoff between the achievable compression rate and access performance: in general, better compression tends to occur by grouping large sets of nodes together before compression, yet such a strategy will often hurt access time by increasing the number of decompression operations needed to extract relevant document fragments.

In this paper, we focus on the permutation-based approaches, and seek to determine the complexity of determining optimal strategies for *container grouping* and *compression algorithm selection* such that the resulting *compression configuration* maximizes the overall compression gain, while keeping compression and/or decompression time and compression model storage requirements within specified bounds. Arion et al [7] were the first to investigate (albeit informally) the tradeoff between compression rate and query performance, given a set of typical queries, a set of available compression algorithms, and a specific XML database as inputs. We consider a more general setting that captures the problem outlined in [7] as well as additional application domains, including data archiving and data exchange. We provide a complexity analysis indicating that the difficulty of selecting an optimal compression configuration is **NP**-hard, and also describe an approximation algorithm based on a branch-and-bound technique that finds the optimal compression configuration within polynomial time (w.r.t. the document size and the number of available compression algorithms), with the choice of appropriate parameter values.

The paper is structured as follows. Section 2 provides preliminary definitions and a background into the problem. Section 3 investigates the difficulty of choosing an optimal tradeoff between compression gain and query performance. Section 4 describes an approximation algorithm for choosing a near-optimal compression configuration, while Section 5 concludes the paper and outlines our future work.

2   Preliminaries

2.1   XML Data Model

We recall that an XML document can be represented as a rooted, ordered, labeled tree (the *document tree*), in which the leaf nodes correspond to attribute values and text segments (document content), while the interior nodes represent attributes and elements (document structure). According to convention, we distinguish attribute names from elements by prepending the former with '@'. As

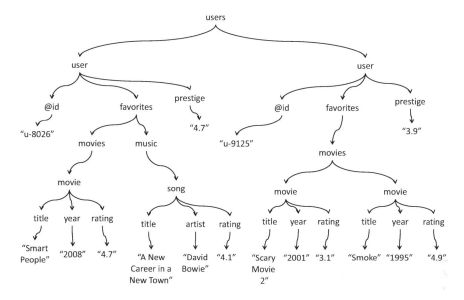

Fig. 1. Example XML document tree

an illustrative example application, we consider a social recommendation website, where users share their opinions of movies, music, etc. with other users. Additionally, users assign a prestige to other users, allowing them to express their evaluation of the quality of those users' recommendations. User account data is stored as XML; Fig. 1 shows a fragment of the document tree.

Query languages for XML center around *path expressions*, which are used to specify subsets of nodes within the document tree. The two most influential XML query languages are XPath [12] and XQuery [13].

Example 1. For the example document tree of Fig. 1, the following XQuery returns the titles of movies rated at least 4.5 by users with a prestige ranking lower than 4.

```
let $movies := for $user in doc(''ratings.xml'')//user
    where $user/prestige lt 4.0
    return $user/favorites/movies/movie

for $movie in $movies
    where $movie/rating ge 4.5
    return $movie/title
```

This query returns `<title>Smoke</title>`.

2.2 XML Compression

Permutation-based strategies for XML-conscious compression separately compress the document structure and text content. The textual content is organized into containers, usually based on the path (or just the name) of the parent element. The intuition for doing so is that values belonging to different instances of the same element are likely to exhibit similarities that facilitate compression. Fig. 2 shows the default path-based partitioning of the text content of the document tree in Fig. 1, in which data values belonging to each distinct element and attribute type stored in a separate container.

Further compression gains can often be realized by generalizing the partitioning strategy to take into account additional factors, such as the data type of the content (e.g., integers, dates, and strings). Grouping together multiple containers with high pairwise similarity allows the containers to share the same compression source model, reducing storage costs while simultaneously allowing more complex models over the longer sequence to be built. Fig. 3 depicts a logical partitioning strategy that extends the default strategy from Fig. 2. Here, containers B, E, and H are grouped together, since user prestige, movie ratings, and song ratings are highly similar (i.e., they all consist of a real number value in the range [0.0, 5.0]). Similarly, since the titles of movies and songs and artist names are all free-form text, it may prove beneficial to group together containers C, F, and G.

The choice of a partitioning strategy can also impact the efficiency of random access to nodes within the document tree. In particular, query performance can be improved by choosing a partitioning strategy that places data segments involved in a common query within the same container subset. Doing so can dramatically reduce the number of required decompression operations. For Ex. 1, a beneficial partitioning strategy might instead group together containers B, C, and E.

Proper algorithm selection is also an important factor to consider. Greater compression can be realized by choosing a compression algorithm that is well-suited for the type of data values stored in a container subset. Query performance is also impacted by the choice of compression algorithm, as the time required to carry out decompression adds to the query response time. Fig. 3 additionally assigns a compression algorithm to each container subset (in this case, either LZ77 or Huffman coding).

Furthermore, certain compression algorithms allow classes of operations to be carried out without prior decompression; the choice of such an algorithm can therefore speed up query performance. For example, using an order-preserving algorithm to compress user prestige and movie rating values would allow the comparisons in both **where** clauses of the XQuery in Ex. 1 to be computed within the compressed domain, without requiring the decompression of each such value beforehand.

We now consider the relevant measures used to evaluate solutions to XML compression problems. *Storage gain* measures the relative amount of space saved by applying a compression algorithm a to a container C, denoted as $gain(C, a)$. An effective measure must not only account for the size of the compressed representation of C; it must also consider the additional space required to store

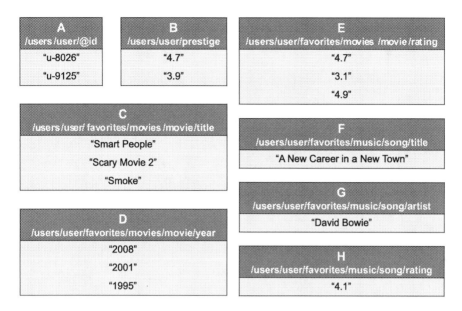

Fig. 2. A path-based partitioning of data values from the document of Fig. 1

auxilary data structures constructed by the compression source model (e.g., for the Huffman algorithm, this would indicate the size of the generated tree; in dictionary-based compression schemes, it would represent the size of the dictionary). It is calculated as

$$gain(C, a) = 1 - \frac{\text{compressed size of } C + \text{compression model size}}{\text{original size of } C} . \tag{1}$$

This measurement is also applicable to *sets* of containers; given a subset $S \subseteq \mathcal{C}$ and a compression algorithm a, $gain(S, a)$ is calculated by first concatenating the contents of each container in S, and then using the compressed and original sizes of this concatenated container, together with the storage costs of the generated compression model, in the above formula.

Compression cost and *decompression cost* measure, respectively, the time required to apply and reverse the compression process. Both time measures are largely dependent on the contents of the container(s) being compressed, as well as the compression algorithm being employed. By $comp(S, a)$ and $decomp(S, a)$, we denote, respectively, the time required to compress and decompress the contents of a container subset S that has been compressed with algorithm a.

In the sequel, we assume that all three measures can be calculated in polynomial time (with respect to the size of the input container subset).

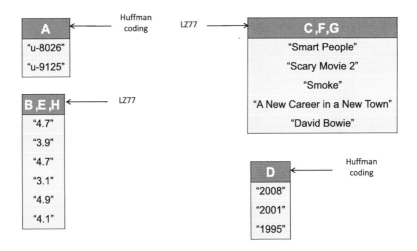

Fig. 3. A compression configuration for the document in Fig. 1

3   Complexity Analysis of Compression Configuration Selection

We recall from the discussion in Sec. 2.2 that our goal is to discover an optimal *compression configuration*, specifying both a partitioning strategy of the container set $\mathcal{C}$ and an assignment of a compression algorithm to each partition set. In this section, we demonstrate the **NP**-hardness of this problem.

Definition 1. *A configuration $\langle P, \alpha \rangle$ consists of a partition $P = \{S_1, \ldots, S_t\}$ of $\mathcal{C}$, and an algorithm assignment function $\alpha : P \to \mathcal{A}$ that assigns to each $S \in P$ a compression algorithm $a \in \mathcal{A}$.*

Definition 2. *An instance of the optimization version of the* optimal compression configuration *problem consists of the following inputs: a set of available compression algorithms $\mathcal{A} = \{a_1, \ldots, a_q\}$; a set of containers $\mathcal{C} = \{C_1, \ldots, C_x\}$; $gain : 2^{\mathcal{C}} \times \mathcal{A} \to \mathbb{Q}$, a function indicating the compression gain obtained when a specific compression algorithm in $\mathcal{A}$ is applied to a specific container subset in $2^{\mathcal{C}}$; $comp : 2^{\mathcal{C}} \times \mathcal{A} \to \mathbb{Q}$, a function indicating the time cost associated with a compression of a specific container subset in $2^{\mathcal{C}}$ using a specific algorithm in $\mathcal{A}$; $decomp : 2^{\mathcal{C}} \times \mathcal{A} \to \mathbb{Q}$, a function indicating the time cost associated with decompressing a specific container subset in $2^{\mathcal{C}}$ that has previously been compressed using a specific algorithm in $\mathcal{A}$; T_c, an upper bound on total compression cost; and T_d, an upper bound on total decompression cost.*

The goal is to discover a configuration $\langle P, \alpha \rangle$ that maximizes $\sum_{S \in P} gain(S, \alpha(S))$ subject to the constraints $\sum_{S \in P} comp(S, \alpha(S)) \leq T_c$ and $\sum_{S \in P} decomp(S, \alpha(S)) \leq T_d$.

In the decision version of the problem, there is an additional input $L \in \mathbb{Q}^*$ and a solver outputs "yes" if there exists a configuration $\langle P, \alpha \rangle$ such that $\sum_{S \in P} gain(S, \alpha(S)) \geq L$ subject to the given constraints, and "no" otherwise.

Theorem 1. Selecting an optimal compression configuration is **NP**-hard.

The proof of Thm. 1 (given in [14]) immediately leads to the following additional result.

Corollary 1. Selection of an optimal compression configuration remains **NP**-hard when $|\mathcal{A}| = 1$.

This indicates that the "hardness" of the overall problem is not caused by algorithm selection, rather it is due to the difficulty of determining an optimal container partitioning strategy.

4   An Approximation Algorithm for Compression Configuration Selection

In this section, we describe an approximation algorithm for selecting an optimal compression configuration. Throughout the discussion, we use the term *container subset* to refer to one or more containers which have been grouped together, and *grouping* to indicate a set of container subsets. A grouping which covers *all* containers (i.e., assigns each container to exactly one container subset) is referred to as a *partitioning strategy*.

In the first phase of the approximation algorithm (Sec. 4.4), a branch-and-bound strategy is used to select a set of *candidate partitioning strategies*: a set of partitioning strategies which are estimated to be highly compressible. In the second phase (Sec. 4.5), this set of partitioning strategies is tested against the set of available compression algorithms to determine the single compression configuration that yields the highest compression gain, while obeying the specified upper bounds on compression and decompression costs.

We first describe how container compressibility and storage costs are estimated, and then discuss how these estimates are used in the computation of compression gains. We then detail both phases of the approximation algorithm.

4.1   Estimating Compressibility

As a means of estimating the compressibility of a container's contents (or of the concatenated contents of multiple containers), we turn to Lempel and Ziv's method for calculating string complexity [15]. In this approach, which we refer to as LZ76, the input string x is parsed once from left-to-right, and a set of phrases $\mathcal{P}_x$ are recursively built and added to a dictionary. Once parsing has been completed, the complexity of x is

$$C_{\text{LZ}}(x) = \frac{|\mathcal{P}_x|}{|x|} , \tag{2}$$

the ratio of phrases per character. Lempel and Ziv showed that this approach yields an approximation ratio of $\frac{n}{\log n}$ to Shannon's entropy rate.

We now describe the parsing process of LZ76 in greater detail. (1) Initialize the dictionary to be empty. (2) If the end of x has been reached, terminate. Otherwise, read the next character from x and assign it to phrase p. If p matches an existing entry in the dictionary, continue reading characters from x and appending them to p until p no longer matches an existing dictionary entry. (3) Assign p the next available index position, and add both the index value and p to the dictionary. Return to step (2).

Example 2. For a container subset S with contents "aaabc", the generated LZ76 dictionary will contain the four phrases $\langle a \rangle$, $\langle aa \rangle$, $\langle b \rangle$, and $\langle c \rangle$ and $C_{LZ}(S) = 4/5 = 0.8$.

4.2   Estimating Storage Cost

To compute the storage gain for a container subset, we simulate the cost of transmitting the dictionary using the coding strategy of LZ78 [16] (recalling that LZ78 utilizes the parsing strategy of LZ76 in concert with a specific coding strategy for dictionary phrases). Each time a new phrase of length l is constructed, two pieces of information are emitted to the compression stream: (1) a codeword W, representing the index position of the existing phrase p of length $l - 1$ that forms a prefix of the new phrase, and (2) the "innovative" character c that is appended to p to form the new phrase. Since phrase indexing begins at 1, the highest index value for a dictionary with t phrases will be t. Using a fixed-length encoding, then, we can express each W value using $\log_2(t)$ bits, requiring a total of $t \cdot \log_2(t)$ bits to encode all codewords. Furthermore, a single character c is emitted each time a new phrase is created, requiring an extra $8 \cdot t$ bits (here, we assume a text encoding that requires a single byte per character is in use; multibyte formats can be incorporated by replacing 8 with the number of bits per character used in the chosen encoding format).

Definition 3. *The* storage cost *(expressed in bits) associated with a container subset S is calculated as*

$$storageCost(S) = t \cdot (8 + \log_2(t)) \tag{3}$$

where t is the total number of entries in the dictionary after an LZ76 parsing of S.

The storage cost (expressed in bits) associated with a container grouping G is calculated as

$$storageCost(G) = \sum_{S \in G} storageCost(S). \tag{4}$$

4.3   Modeling Compression Gain

Two distinct gain measures are associated with each container grouping: the *local compression gain (localGain)* indicates the compression gain obtained by using the current grouping, while the *maximum potential compression gain (mpGain)* indicates the highest possible compression gain that can be obtained moving forward by chosing any partitioning strategy that "agrees with" the current grouping (i.e., there exists no container C such that the current grouping and the partitioning strategy place C within different container subsets). Both measures are used in the first phase of the algorithm to guide the search for candidate container partitioning strategies, and we presently describe how both measures are calculated.

Definition 4. *The* local compression gain *(expressed in bits) of a container subset S, denoted* localGain(S)*, is calculated as*

$$localGain(S) = \max\{0, \Gamma(S)\} \ , \tag{5}$$

where

$$\Gamma(S) = 8 \cdot |S| - (C_{\mathrm{LZ}}(S) \cdot |S| + storageCost(S)) \tag{6}$$

and $|S|$ indicates the total byte length of the contents of S.

Eq. (5) ensures that compression is only applied if it results in a positive compression gain; otherwise, the subset S is left *uncompressed*, and $localGain(S) = 0$. In Eq. (6), the sum of the estimated compressed size of S and the associated storage cost is subtracted from the original bit length of S. This quantity represents the total number of bits saved by applying compression to S. Note that while Eq. (6) assumes a byte-level compression of container contents, text encoding schemes using multiple bytes per character (e.g., Unicode formats) may be supported by considering each byte as an individual token.

Definition 5. *The* local compression gain *(expressed in bits) of a container grouping G is calculated as*

$$localGain(G) = \sum_{S \in G} localGain(S) \ . \tag{7}$$

Example 3. Recalling the example grouping $S = \{aaabc\}$ from Ex. 2, $\Gamma(S) = 5 \cdot 8 - (0.8 \cdot 5 + (4 \cdot (8 + \log 2(4)))) = -4$ bits and therefore $localGain(S) = 0$ bits, indicating that S should be left uncompressed.

As mentioned above, the maximum potential compression gain is used to indicate the upper bound on the achievable compression gain for any partitioning strategy that agrees with the current grouping. Since the total number of characters (i.e., the number of characters contained within the existing grouping G, plus the number of unprocessed characters contained within containers that have yet to be assigned to subsets) is fixed, so too is the first product in Eq. (6), and

Input: D, the set of existing LZ76 dictionaries for the grouping G; c_t, total number of characters in all containers of $\mathcal{C}$; c_u, number of remaining unprocessed characters.

Output: $mpGain(G)$, indicating the maximum potential compression gain for G.

1. Choose the dictionary $d \in D$ containing the phrase of longest length, and let S_{max} denote the container subset whose dictionary is d. In case of a tie, choose the subset with the lowest C_{LZ} value. Set $nPhrases$ to be the number of phrase entries in d, and $maxPhraseLength$ to be the length of the longest phrase, plus one.
2. While $c_u \geq maxPhraseLength$, simulate the creation of a new, longer phrase by performing the following steps:
   (a) Set $c_u = c_u - maxPhraseLength$.
   (b) Set $nPhrases = nPhrases + 1$.
   (c) Set $maxPhraseLength = maxPhraseLength + 1$.
3. If $c_u > 0$, choose an existing phrase of length c_u from d to cover the remaining unprocessed characters.
4. Compute $C_{\mathrm{LZ}}(S_{max}) = \frac{nPhrases}{c_t}$, and use this value to recalculate $localGain(S_{max})$.
5. Return $mpGain(G) = \sum_{S \in G \backslash S_{max}} localGain(S) + localGain(S_{max})$.

Algorithm 1. Calculation of maximum potential compression gain

maximizing compression gain over a subset S then requires the sum of $C_{\mathrm{LZ}}(S)$ and $storageCost(S)$ to be minimized. From Eq. (2) and Eq. (3), one observes that both quantities are minimized when the number of generated phrases is also minimized. Equivalently, at each step during LZ76 parsing, one seeks to generate the *longest applicable phrase* by appending an extra character to the longest existing phrase in the dictionary. Alg. 1 illustrates how the maximum potential gain is calculated for a grouping.

In the first step, the longest phrase over all subset dictionaries is identified. For the container subset S_{max} whose dictionary contains this longest phrase, the existing dictionary is extended with longer phrases, until no unprocessed characters remain. More precisely, each iteration of step 2 creates a new phrase one character longer than the previous longest phrase (as we are free to assign arbitrary values to unprocessed characters, such a phrase can always be constructed), and applies it to the sequence of unprocessed characters. Eventually, either all remaining characters will be processed, or the number of remaining characters will be less than the longest phrase. In the latter case, a shorter existing phrase is reused to cover the remaining characters (step 3). Step 4 computes the new value of $C_{\mathrm{LZ}}(S_{max})$, and updates the value of $localGain(S_{max})$. Finally, step 5 computes the $mpGain$ for the grouping G (expressed in bits) by summing the updated $localGain$ score for S_{max} with the existing $localGain$ scores for the remaining subsets in G.

Example 4. To illustrate the computation of $mpGain$, we recall from Ex. 2 the previous example subset $S = \{aaabc\}$, and the dictionary of phrases

$\{\langle a \rangle, \langle aa \rangle, \langle b \rangle, \langle c \rangle\}$ that results from an LZ76 parsing of S. Assume that there is one additional container C_x with 5 characters. Alg. 1 first selects the longest existing phrase $\langle aa \rangle$ and constructs a new phrase of length 3 (say, $\langle aaa \rangle$). Applying this to C_x leaves only $5 - 3 = 2$ remaining unprocessed characters, a number which is less than 3, the current maximum phrase length. Therefore, the existing pattern $\langle aa \rangle$ is applied, and no unprocessed characters remain. Only one additional pattern has been created, and the new complexity score is $5/10 = 0.5$ symbols per character, while the updated storage cost is $5 \cdot (8 + \log_2(5)) \approx 51.6096$ bits, and $mpGain(S) \approx 10 \cdot 8 - (0.5 \cdot 10 + 51.6096) \approx 23.3904$ bits.

4.4 Branch-and-Bound Algorithm for Selecting Candidate Partitioning Strategies

In this phase, a search tree is constructed in which each node corresponds to a particular grouping. Each node stores the *localGain* and *mpGain* values for its associated grouping. The subtree rooted by a node n encompasses all groupings that extend the grouping associated with n by assigning additional containers to container subsets.

Before explaining the details of the branch-and-bound procedure, we begin with an intuition as to why this technique is applicable to the subproblem of choosing a container grouping. Recall that the *mpGain* indicates the highest possible gain possible for any partitioning strategy based on the current grouping. In addition, we may also observe that $mpGain(p) \geq mpGain(c)$ for any parent node p and child node c in the search tree. This is due to the fact that there are fewer remaining unprocessed characters as one travels from p to c: in particular, the placement of one additional container has been "fixed" by the grouping associated with c. At the lowest level of the search tree, *all* containers have a fixed placement (i.e., each leaf node corresponds to a partitioning strategy), and therefore *mpGain* will equal *localGain* for each leaf node.

Exploiting these properties of the *mpGain* measure provides us with our *bounding criterion*: if the *mpGain* for a grouping is sufficiently less than the best local gain value encountered thus far, the entire subtree rooted at the node representing the grouping can be immediately eliminated from consideration (or "killed"). We now are in a position to describe the specifics of the branch-and-bound procedure.

The inputs to the procedure are a set of containers $\mathcal{C}$, sorted in descending order of their respective sizes, along with an additional parameter $\delta \in \mathbb{R}^+$. The latter specifies a threshold value used to determine whether a particular node should be "killed", or if it is worthwhile to continue branching into its subtree (in which case it is considered to be a "live" node). During the search procedure, the optimal local gain value encountered so far is stored in variable *optGain*. The root node of the search tree is assigned the grouping $\{C_1\}$, that is, a single set containing only the first container. For $i = 2, ..., |\mathcal{C}|$, the steps in Alg. 2 are carried out to enumerate the various choices for placement of each container

Input: container $C_i \in \mathcal{C}$, context node x, and a threshold value $\delta \in \mathbb{R}^+$
Output: a set $\mathcal{G}$ of candidate container partitioning strategies
1. Construct as the leftmost child of x the grouping formed by adding the single-container subset $\{C_i\}$ to G_x.
2. For each existing subset $S \in G_x$, add a child to x corresponding to the grouping formed by $G_x \setminus \{S\} \cup \{S \cup C_i\}$.
3. For each of the child nodes y created in steps 1 and 2, let G_y represent the grouping associated with y and calculate $localGain(G_y)$ and $mpGain(G_y)$.
4. If one of the newly constructed children nodes y results in a $localGain(G_y)$ value that is higher than $optGain$, set $optGain$ to this value.
5. "Kill" any child nodes y for which $mpGain(G_y) < optGain - \delta$.

Algorithm 2. Construction of the branch-and-bound search tree

C_i within the context of an existing grouping (where each such choice corresponds to a child node of the existing grouping node), and to determine the optimal choice of placement among the alternatives. Note that in Alg. 2, x refers to the node currently being evaluated in the tree, and G_x refers to the container grouping associated with x.

For each "live" node p at level i in the tree, a set of child nodes are constructed; each represents a different strategy for placing the container C_{i+1} into either a new subset, or within one of the existing container subsets present in the grouping associated with p. Once all "live" nodes at level i have been branched, $mpGain$ and $localGain$ values for all nodes at level i are computed, and if necessary $optGain$ is updated to reflect a new global maximum for $localGain$. For each node having a $localGain$ less than $optGain$, a test is carried out to ensure that its $mpGain$ falls within the range $[optGain - \delta, optGain]$. If the test fails, the node is "killed". Further branching is only carried out at level $i+1$ on the remaining live nodes at level i; at each iteration, the unbranched node at level i with the highest $mpGain$ value is chosen. At level $|\mathcal{C}|$, the remaining live nodes will comprise the set $\mathcal{G}$ of candidate partitioning strategies.

We illustrate the working of the branch-and-bound procedure with the following example.

Example 5. Assume that we have the container set $\mathcal{C} = \{C_1, C_2, C_3\}$, where the respective contents of the containers are $C_1 = \{aaabcaaabcaaabcabcab\}$, $C_2 = \{15720653197608243849\}$, and $C_3 = \{abcababcbaaaabcabcab\}$. We set $\delta = 30.0$ bits. Fig. 4 depicts the search tree formed by this process. The best local gain is achieved by the grouping $\{C_1, C_3\}, \{C_2\}$; when we compare the $mpGains$ of the other nodes at level 3, only $\{C_1\}, \{C_2\}, \{C_3\}$ comes within $\delta = 30.0$ bits of this optimal local gain. Hence, only these two nodes remain alive, and the other three are "killed". Since all three containers have now been assigned, we return the two remaining live nodes at level three as the set of candidate partitioning strategies, $\mathcal{G}$.

The pruning criterion in step 5 serves to reduce the size of the search space, yet it is crucial to ensure that it does not result in the removal of the node with the

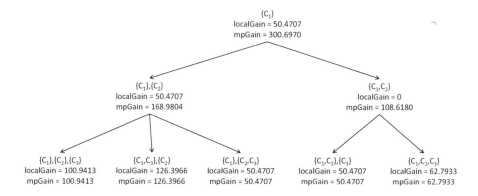

Fig. 4. Branch-and-bound search tree for Ex. 5

highest local compression gain (the optimal node). The following result proves that the optimum node will never be "killed".

Proposition 1. *Alg. 2 ensures that the optimal node is always visited.*

Proof. Given in [14].

4.5 Determining an Optimal Compression Configuration

Alg. 3 allows one to determine an optimal compression configuration from an input set $\mathcal{G}$ of candidate partitioning strategies (obtained from Alg. 2) and set $\mathcal{A}$ of compression algorithms, together with upper bounds on compression and decompression time, T_c and T_d. The variable *globalBestGain* records the highest overall compression gain from the partitioning strategy/algorithm assignment combinations tested so far. Lines 2-29 iterate through each candidate partitioning strategy $G \in \mathcal{G}$; each container subset S contained in G is tested (Lines 4-23) to determine the compression algorithm $a \in \mathcal{A}$ that achieves the highest compression gain (Lines 6-14). Before an algorithm is assigned to a container subset, a test is performed to ensure that the required compression and decompression time values fall below the respective bounds T_c and T_d (Line 8).

At the conclusion of testing, if there is no available algorithm in $\mathcal{A}$ that satisfies the time bounds for compression and decompression when applied to a specific subset S, the entire partitioning strategy containing S is immediately disqualified (Lines 15-16). Otherwise, the overall compression gain and compression/decompression time scores are updated for the partitioning strategy G, and the appropriate compression algorithm is assigned to the active subset S (Lines 17-22). After each partitioning strategy G has been processed, a test is done to determine whether it yields a better gain than the current *globalBestGain*; if necessary, the globally-best compression configuration $\langle P, arg \rangle$ is updated to store the current partitioning strategy G, along with the optimal algorithm selection strategy α_G found for G (Lines 24-28).

Input: set of compression algorithms $\mathcal{A}$, set of candidate container partitions $\mathcal{G}$, upper bound $T_c \in \mathbb{Z}^+$ on compression time, upper bound $T_d \in \mathbb{Z}^+$ on decompression time

Output: a compression configuration $\langle P, \alpha \rangle$

1 $globalBestGain \leftarrow 0$; $P \leftarrow NULL$; $alg \leftarrow NULL$;
2 **foreach** $G \in \mathcal{G}$ **do**
3      $groupingCTime \leftarrow 0$; $groupingDTime \leftarrow 0$; $groupingGain \leftarrow 0$;
4   **foreach** $S \in G$ **do**
5          $maxGain \leftarrow 0$; $bestCTime \leftarrow 0$; $bestDTime \leftarrow 0$;
            $bestAlgorithm \leftarrow NULL$;
6   **foreach** $a \in \mathcal{A}$ **do**
7              $gain \leftarrow compressedGain(S, a)$;
8   **if** $gain > maxGain$ **and**
                $groupingCTime + compressTime(S, a) \leq T_c$ **and**
                $groupingDTime + decompressTime(S, a) \leq T_d$ **then**
9                  $bestCTime \leftarrow compressTime(S, a)$;
10                 $bestDTime \leftarrow decompressTime(S, a)$;
11                 $bestAlgorithm \leftarrow a$;
12                 $maxGain \leftarrow gain$;
13   **end**
14   **end**
15   **if** $bestAlgorithm = NULL$ **then**
16   **goto** line 4;
17   **else**
18             $groupingCTime \leftarrow groupingCTime + bestCTime$;
19             $groupingDTime \leftarrow groupingDTime + bestDTime$;
20             $groupingGain \leftarrow groupingGain + maxGain$;
21             $\alpha_G(S) \leftarrow bestAlgorithm$;
22   **end**
23   **end**
24   **if** $groupingGain > globalBestGain$ **then**
25         $globalBestGain \leftarrow groupingGain$;
26         $P \leftarrow G$;
27         $\alpha \leftarrow \alpha_G$;
28   **end**
29 **end**
30 **return** $\langle P, \alpha \rangle$;

Algorithm 3. Selecting a compression configuration

After all partitions in $\mathcal{G}$ have been processed, the optimal compression configuration $\langle P, \alpha \rangle$ is returned (Line 30).

5   Conclusion

In this paper, we demonstrated that determining an optimal configuration for permutation-based XML compression is an **NP**-hard problem. We also described an approximation algorithm that allows one, with proper selection of parameter values, to discover the optimal compression configuration in polynomial time (w.r.t. the sizes of the document and the set of compression algorithms $\mathcal{A}$). As future work, we plan to implement this algorithm within our existing XML-conscious compressor [8] and test its effectiveness via experimentation over a range of real-world and synthetic XML documents.

References

1. Adiego, J., la Fuente, P.D., Navarro, G.: Combining structural and textual contexts for compressing semistructured databases. In: ENC, pp. 68–73 (2005)
2. Cheney, J.: Compressing XML with multiplexed hierarchical PPM models. In: DCC, pp. 163–172 (2001)
3. Cheney, J.: An empirical evaluation of simple DTD-conscious compression techniques. In: WebDB, pp. 43–48 (2005)
4. Leighton, G., Müldner, T., Diamond, J.: TREECHOP: a tree-based query-able compressor for XML. In: CWIT, pp. 115–118 (2005)
5. Min, J.K., Park, M.J., Chung, C.W.: XPRESS: A queriable compression for XML data. In: SIGMOD, pp. 122–133 (2003)
6. Tolani, P.M., Haritsa, J.R.: XGRIND: A query-friendly XML compressor. In: ICDE, pp. 225–234 (2002)
7. Arion, A., Bonifati, A., Manolescu, I., Pugliese, A.: XQueC: A query-conscious compressed XML database. ACM TOIT 7(2), Article 10 (May 2007)
8. Leighton, G., Diamond, J., Müldner, T.: AXECHOP: a grammar-based compressor for XML. In: DCC, p. 467 (2005)
9. Liefke, H., Suciu, D.: XMill: An efficient compressor for XML data. In: SIGMOD, pp. 153–164 (2000)
10. Maneth, S., Mihaylov, N., Sakr, S.: XML tree structure compression. In: XANTEC, pp. 243–247 (2008)
11. Skibinski, P., Grabowski, S., Swacha, J.: Effective asymmetric XML compression. Software: Practice and Experience 38(10), 1027–1047 (2008)
12. Berglund, A., Boag, S., Chamberlin, D., Fernández, M.F., Kay, M., Robie, J., Siméon, J. (eds.): XML path language (XPath) 2.0, W3C Recommendation (January 2007), http://www.w3.org/TR/xpath20/
13. Boag, S., Chamberlin, D., Fernández, M.F., Florescu, D., Robie, J., Siméon, J. (eds.): XQuery 1.0: An XML query language, W3C Recommendation (January 2007), http://www.w3.org/TR/xquery/
14. Leighton, G., Barbosa, D.: Optimizing XML compression (extended version). CoRR abs/0905.4761 (2009), http://arxiv.org/abs/0905.4761
15. Lempel, A., Ziv, J.: On the complexity of finite sequences. IEEE Trans. Inf. Theory 22(1), 75–81 (1976)
16. Ziv, J., Lempel, A.: Compression of individual sequences via variable-rate coding. IEEE Trans. Inf. Theory 24(5), 530–536 (1978)

XML Lossy Text Compression:
A Preliminary Study

Angela Bonifati[1,2], Marianna Lorusso[2], and Domenica Sileo[2]

[1] Italian National Research Council (CNR)
Via P. Bucci 41C, I-87036 Rende, Italy
`bonifati@icar.cnr.it`
[2] Dipartimento di Matematica e Informatica, University of Basilicata
Viale dell'Ateneo Lucano 10, I-85100 Potenza, Italy
`marianna.lorusso@gmail.com`, `domenica.sileo@gmail.com`

Abstract. Lossy compression techniques have been applied to image and text compression, yielding compression factors that are vastly superior to lossless compression schemes. In this paper, we present a preliminary study on a set of lossy transformations for XML documents that preserve the semantics. Inspired by previous techniques, e.g. lossy text compression and literate programming, we apply a simple algorithm to XML syntactic constructs to loose superfluous layout information and redundant text. The obtained XML keeps the human-readability and machine-readability properties. Additionally, it can lead to a considerable reduction of its space occupancy and boost the application of conventional text compressors, thus representing a promising technology for several data management tasks.

1   Introduction

Lossy compression leads to produce compressed files that cannot be reconstructed in their original form. Such compression can be used alone or in conjunction with lossless compression to improve the compression rate. We focus on lossy compression techniques to be applied to synthetic languages and to natural language as well. An XML file is indeed a combination of both languages, as it consists of syntactic constructs, such as elements, attributes, comments, entities, processing instructions etc. and of natural text, which is embedded into PCDATA nodes. As such, XML can be shrunk by eliminating layout information, such as whitespaces and closing tags. Such techniques have been devised for natural languages in lossy text compression [1], where thesauri-based compression is employed, by replacing words with their shorter synonyms. In synthetic languages, the lossy text compression intends to compress the text but preserve the semantics. For instance, source code can be compressed by eliminating superfluous white space. The obtained program will not be human-readable as it used to be, but will still be accepted by its compiler. Not only layout information but also comments are eliminated from a source program in WEB [2], to obtain a 'tangled' version of the language, which is not intended for human consumption but only for compilers consumption. Literate programming aims at creating

Z. Bellahsène et al. (Eds.): XSym 2009, LNCS 5679, pp. 106–113, 2009.
© Springer-Verlag Berlin Heidelberg 2009

two versions of a given program, a compiler version and a pretty-printing tex version, the latter being suitable as code documentation. Finally, the common technique of omitting vowels from text, while being hardly suitable for practical use, sacrifices readability for compression. Its variations include Speedwriting [3], Braille [4] and Soundex [5].

All the aforementioned techniques are not directly applicable to XML syntactic constructs and need to be customized. XML has a double-fold nature, in that it has instructions (i.e. the components of the data model) and it also encloses textual data. These data typically reside in the leaves and can be quite large for full-text documents. Lossy compression of natural language has been tackled in the past, by yielding outputs seldom readable by humans. Whereas these solutions were acceptable for text, they are not viable for XML data, that has to keep its human-readability and machine-readability at any rate.

Our first contribution has been that of devising a set of rules for lossy text compression of an XML file. We have identified a set of rules applicable to the syntactical constructs of an XML file, a set of rules for compressing its textual content, and a set of rules to reduce its formatting content. We obtain two variants of an XML file, that are both machine-readable, the first preserving the well-formedness property, and the second sacrificing it for compression. We call these variants $\mathcal{WF}$-Lossy XML and $\mathcal{TF}$-Lossy XML ($\mathcal{WF}$-LX and $\mathcal{TF}$-LX, in short notation). Before explaining the acronyms, we observe that, while the first of the two variants is still an XML file, thus can be processed as such, both variants do keep the human-readability property. $\mathcal{WF}$-LX files represent well-formed Lossy XML files, whereas $\mathcal{TF}$-LX represent text-formed Lossy XML files, i.e. files that can be read as text, although their conversion to a well-formed document is straightforward (cfr. Section 2). *Our second contribution has been that of implementing a Lossy XML Compressor (LXC), capable of yielding both formats, and studying its effectiveness on several XML datasets.* From our analysis, we observed that our rules let achieve a moderate compression factor in some of the considered datasets, and a significant reduction (up to 40%) in a few other datasets. *Our third contribution is that of identifying a class of applications, in order to show the utility of our approach.* It does appear that some very interesting work can be done in this area.

The rest of the paper is organized as follows. Section 2 describes the rules to obtain lossy XML files. Section 3 shows a set of experiments to gauge the effectiveness of our technique and its usage in conjunction with ordinary compressors. Section 4 discusses the related work. Finally, Section 5 concludes our paper, and discusses future directions of our work.

2   Rules for XML Lossy Text Compression

We describe in the following the rules we have devised to compress the textual content of an XML file in a lossy fashion. These rules are divided into three main categories: *(i)* PCData rules, i.e. only applicable to the leaves of an XML tree, being such leaves the PCData values of attributes or elements; *(ii)* Tag

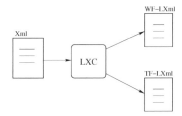

Fig. 1. Lossy XML Compression (LXC) yields two output files

name rules, i.e. only applicable to intermediate nodes of an XML tree, i.e. to the starting and closing tags of an element, to attribute names and to namespace nodes; *(iii)* Formatting rules, i.e. applicable to various formatting characters that lie in between nodes in an XML tree.

PCData Rules. The rules for PCDATA are as follows:

1. a sequence of one or more vowels is replaced with one vowel, typically the first;
2. a sequence of one or more punctuation characters is replaced with the empty sequence if there is a blank afterwards, otherwise with a whitespace; this rule is not applied in cases where the punctuation is followed by a number, or when the word begins with the character '&' and ends with the character ';' as this represents an entity;
3. a sequence of two or more formatting characters is replaced with one formatting character;
4. 's and s' (genitive inflection) are eliminated from words;
5. a word duplicate if it appears after the first duplicate is eliminated;
6. numbers written in letters are replaced with the corresponding digits;
7. whitespaces in the leading sentence of a given paragraph and whitespaces in the trailing sentence of a paragraph are deleted;
8. if there is an acronym followed by its full expansion, the latter is eliminated; any full expansion in the document is replaced by the acronym;
9. end of line characters are removed and replaced by a whitespace;
10. if a word that appears after the character '.', '!' and '?' begins with a low-ercase letter, the letter is converted in uppercase.

Formatting Rules. The rules for formatting are as follows:

1. end of line characters and lines of whitespaces are eliminated;
2. comments are removed;
3. indentation characters are removed;
4. sequences of two or more formatting characters are replaced with one formatting characters, typically the first;
5. leading and trailing whitespaces are removed.

Table 1. Prioritized order among rules - table must be read *per row*; TR (PR, FR, resp.) stands for Tag Rule (PCData Rule, Formatting Rule, resp.)

FR3	FR2	*TR3*	FR5	PR8	PR4
PR5	PR10	PR2	PR6	FR4	PR3
PR1	PR7	TR2	TR1	PR9	FR1

Tag Rules. The rules for tag names, attributes and namespaces are as follows:

1. sequences of two or more vowels are replaced with one vowel, typically the first;
2. characters '. ', '-' and '_' are removed;
3. each closing tag name is eliminated, and left as the acute brackets with the character '/' (i.e. `</>`).

We have prioritized the rules both within each class and across distinct classes. This boils down to decide a global ordering for rules, which can be read per row in Table 1. To give an intuition, rules need to be prioritized in order to avoid redundant work (e.g. PCData rule 6 is applied before PCData rule 1, in order to avoid collapsing consecutive vowels in numbers) and to guarantee that the application of a rule is not void (e.g. PCData rule 10 is fired before PCData rule 2, in order to avoid loosing the separation into sentences before removing redundant punctuation characters). All rules must be applied to obtain a $\mathcal{WF}$-LX file. All rules but Tag Rule 3 are applied to obtain a $\mathcal{TF}$-LX file.

Before discussing the implementation of our prototype, we would like to underscore the importance of having Tag Rule *TR3*, which substitutes the final closing tag of an element with its empty version `</>`. If such substitution takes place, the obtained $\mathcal{TF}$-LX document looses its well-formedness. However, such document, as we will see from the experimental result, becomes more apt to compression. Notwithstanding the advantages of keeping the document in textual format, one can always reconstruct the XML closing tags and transform every $\mathcal{TF}$-LX document into an $\mathcal{WF}$-LX document. We also observe that a depth-first encoding can be applied to $\mathcal{TF}$-LX documents, similarly to that applied to n-ary trees or balanced mathematical expressions [6]. Indeed, closing empty tags act as placeholders, and can be easily matched to opening tags during document reconstruction. As such, closing tags resemble closing parentheses in mathematical expressions. An balanced parentheses encoding [6] can be applied to the obtained lossy $\mathcal{TF}$-LX document, which we plan to investigate as future work.

3   Experimental Study

We have conducted an experimental study in order to gauge the effectiveness of our technique and the succinctness of the obtained documents. In particular,

Table 2. XML documents used

Document d	Size (KB)	# Elems.	# Attributes	Max Depth	Provenance
Path	203	2764	8627	10	[7]
XMark	113,794	1,666,315	381,878	13	[8]
DBLP	932,444	16,272,139	14,936,399	7	[9]
Shakespeare	274	6636	0	8	[10]
TreeBank	84,065	2,437,666	1	37	[9]
SwissProt	112,130	2,977,031	2,189,859	6	[9]
News	238,677	3,974,681	0	3	[11]

we have run a first set of experiments, aiming at measuring the compression ratio of both $\mathcal{W}F$-LX documents and $\mathcal{T}F$-LX documents. Then, we have compared these compression factors with existing compressors. Next, we have run a second set of experiments, by sequencing the application of our lossy compression technique and of general-purpose compressors. Finally, as a third set of experiments, we have loaded our $\mathcal{W}F$-LX documents into an XML database engine and measured both the loading time and the query execution times.

3.1   Experimental Setting and Results

We have performed our experiments on a Windows XP Pro Laptop with 2.40 Ghz Intel Core Duo CPU P8600, and 4 GB RAM. The datasets used in the experimental study are shown in Table 2. Their structure varies from flat (e.g. DBLP) to deeply nested (e.g. XMark and TreeBank).

Figure 2 (top) shows the compression ratio obtained for the above datasets by using our tool LXC. We can observe that the effectiveness of the rules is higher once the textual content of a document is higher, and that the technique performs quite well in some datasets, such as DBLP and TreeBank (up to 47%). Such datasets exhibit both compressible tags and compressible textual content. The remaining datasets can achieve a moderate compression ratio, that goes from 20% to 10%. In Figure 2 (bottom), we compare our results with the compression ratios obtained by two general-purpose compressors, such as GZip and BZip2 [12], two XML compressors, i.e. XMill [13] and Xmlppm [14], and Huffword (word-oriented Huffman [15]). Although our technique is fairly different from the opaque lossless compression obtained by the above tools, we can observe that the compression ratio obtained by our lossy technique can be up to an half of the compression ratio of both classical and XML compressors, and can be comparable to the compression ratio achieved by Huffword on some datasets.

We have next analyzed the impact of our technique on the effectiveness of classical and XML compressors. Hence, we have sequentially applied our technique and classical compressors. We did not observe a significant variation of the compression ratio for GZip, BZip2, XMill and Xmlppm. Indeed, by either

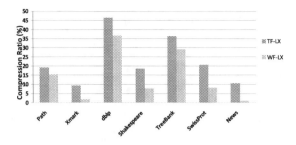

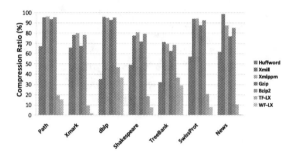

Fig. 2. Compression Ratio (%) for various XML datasets of $\mathcal{T}F$-LX against $\mathcal{W}F$-LX (top); Compression Ratio (%) for various XML datasets of $\mathcal{T}F$-LX and $\mathcal{W}F$-LX (bottom) against other compression tools

applying GZip (BZip2, resp.) to $\mathcal{T}F$-LX files[1], or by applying GZip (BZip2, Xmill, Xmlppm, resp.) to $\mathcal{W}F$-LX files, the compression ratio of these tools is the same of that achieved when working on the original document, or, in some cases, slightly decreases. This demonstrates that the above compressors either ignores the changes applied by our rules, or cannot take advantage of them. Due to the lack of space, we opt not to present the above results and only show the figures relative to Huffword. Indeed, this was the only compressor that could exploit the reduction induced by our technique. Figure 3 shows the results of applying Huffword to the $\mathcal{T}F$-LX files, Huffword to the original files, and Huffword to the $\mathcal{W}F$-LX files, by using the datasets of Table 2. We can observe that in most of the cases, lossy compression boosts Huffword compression ratio, or, at least, does not change it. Only in one case, the compression ratio decreases, and this happens with TreeBank. The motivation behind this is the fact that TreeBank contains partially encrypted data, and these data cannot be processed by our technique. This result lets us thinking that TreeBank compression ratio would have improved if those data were decrypted.

---

[1] We did not employ XMill and Xmlppm in this experiment, as these are not applicable to $\mathcal{T}F$-LX files.

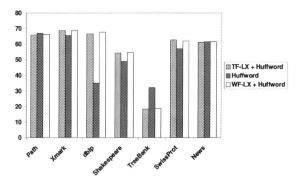

Fig. 3. Compression Ratio (%) for various XML datasets of applying $\mathcal{T}F$-LX and Huffword, Huffword alone, and $\mathcal{W}F$-LX and Huffword

To conclude our experiments, we showed the utility of $\mathcal{W}F$-LX documents into XML databases. We considered QizX [16], a quite fast XML database and XQuery engine, and we uploaded TreeBank and DBLP[2] after applying lossy compression. We observed that the loading time of $\mathcal{W}F$-LX documents was on average 10% less than the loading time of the original documents. Then, we executed a set of XPath queries on both datasets and observed a reduction of 20% of the query execution times on average.

4   Related Work

Techniques for word abbreviations have been the subject of past research on textual data [5], where various word abbreviation techniques are compared, among which Soundex, a phonetic algorithm for indexing names patented by R.C. Russel in 1918. The Soundex indexing system is still in use nowadays by the US Census, to systematically code persons' names. Each word in the Soundex alphabet retained the same degree of discrimination of the original word, along with the mnemonic similarity with the original word. Moreover, the procedure can be applied on the fly, i.e. without any prior knowledge of the population of words, and typically at any new person filed at Census. We observe that the no storage or table lookup is necessary in our technique and the above principles are as well satisfied by our rules. Whereas a large body of research studied the problem of opaque compression for XML files [13,14,17,18], to the best of our knowledge no previous work aimed at envisioning abbreviation techniques for XML constructs and textual content. The only work we are aware of about lossy compression is [19], but their aim is to build a synopsis of the original document, reporting aggregate data, which is quite different from the text abbreviation rules we have presented in this paper.

---

[2] We only used TreeBank and DBLP in these experiments, as they obtained the highest compression ratio under our technique.

5   Conclusions and Future Perspectives

We have described our preliminary study on XML lossy text compression. We believe that there exists several interesting directions of future research. In particular, many challenges are left to be addressed, such as designing a full-text engine for $\mathcal{TF}$-LX files and studying keyword-based queries for such documents. Another important milestone is to devise an extension to XQuery that handles queries on $\mathcal{WF}$-LX files in the compressed domain, as in our past work [17].

References

1. Witten, I.H., Bell, T.C., Moffat, A., Nevill-Manning, C.G., Smith, T.C., Thimbleby, H.W.: Semantic and generative models for lossy text compression. The Computer Journal 37(2), 83–87 (1994)
2. Knuth, D.E.: Literate programming. The Computer Journal 27(2), 97–111 (1984)
3. SpeedWriting, http://www.speedwriting.co.uk/
4. Braille, http://www.nfb.org/nfb/BrailleInitiative.asp
5. Bourne, C.P., Ford, D.F.: A study of methods for systematically abbreviating english words and names. J. ACM 8(4), 538–552 (1961)
6. Benoit, D., Demaine, E.D., Munro, J.I., Raman, R., Raman, V., Rao, S.S.: Representing Trees of Higher Degree. Algorithmica 43(4), 275–292 (2005)
7. PathWays, http://www.genome.jp/kegg/xml
8. Schmidt, A., Waas, F., Kersten, M., Carey, M., Manolescu, I., Busse, R.: XMark: A benchmark for XML data management. In: Proceedings of VLDB, pp. 974–985 (2002)
9. University of Washington's XML repository (2004), www.cs.washington.edu/research/xmldatasets
10. Ibiblio.org web site (2004), www.ibiblio.org/xml/books/biblegold/examples/baseball/
11. AG's corpus of News articles, http://www.di.unipi.it/gulli/newsspace200.xml.bz
12. The bzip2 and libbzip2 Official Home Page (2002), http://sources.redhat.com/bzip2/
13. Liefke, H., Suciu, D.: XMILL: An Efficient Compressor for XML Data. In: Proceedings of the 2000 ACM SIGMOD International Conference on Management of Data, Dallas, TX, USA, pp. 153–164. ACM, New York (2000)
14. Cheney, J.: Compressing XML with Multiplexed Hierarchical PPM Models. In: DCC, pp. 163–172 (2001)
15. Huffman, D.A.: A Method for Construction of Minimum-Redundancy Codes. In: Proc. of the IRE, pp. 1098–1101 (1952)
16. Qizx, http://www.xfra.net/qizxopen/
17. Arion, A., Bonifati, A., Manolescu, I., Pugliese, A.: XQueC: A query-conscious compressed XML database. ACM Trans. Internet Techn. 7(2) (2007)
18. Ng, W., Lam, Y.W., Cheng, J.: Comparative Analysis of XML Compression Technologies. World Wide Web Journal 9(1), 5–33 (2006)
19. Cannataro, M., Carelli, G., Pugliese, A., Saccá, D.: Semantic Lossy Compression of XML Data. In: Proceedings of KRDB (2001)

XQuery Full Text Implementation in BaseX

Christian Grün, Sebastian Gath, Alexander Holupirek, and Marc H. Scholl

Department of Computer & Information Science
University of Konstanz
Box D 188, 78457 Konstanz, Germany
firstname.lastname@uni-konstanz.de

Abstract. BaseX is an early adopter of the upcoming XQuery Full Text Recommendation. This paper presents some of the enhancements made to the XML database to fully support the language extensions. The system's data and index structures are described, and implementation details are given on the XQuery compiler, which supports sequential scanning, index-based, and hybrid processing of full-text queries. Experimental analysis and an insight into visual result presentation of query results conclude the presentation.

1 Introduction

XML has been widely adopted as an exchange and storage format for textual data in both research and industry. The existence of more than fifty XQuery processors clearly underlines the large interest in querying XML documents and collections. While many of the database-driven implementations offer their own extensions to support full-text requests, the upcoming XPath and XQuery Full Text 1.0 Recommendation [1] will satisfy the need for a unified language extension and will most probably attract more developers and users from the Information Retrieval community. The recommendation offers a wide range of content-based query operations, classical retrieval tools such as Stemming and Thesaurus support, and an implementation-defined scoring model that allows developers to adapt their database to a large variety of use-cases and scenarios.

In this paper, we present aspects of the implementation of XQuery Full Text in the database system BaseX [14,15,17]. GalaTex [7] and Quark [4] were two systems that supported early versions of the proposal, and BaseX is, to the best of our knowledge, the first implementation to fully support all features of the specification. More implementations are expected to follow in the near future as soon as the recommendation has reached its final state.

A simple full-text test looks nearly the same as a General Comparison in XQuery [5]. An ftcontains expression can get pretty large, however, if the right-hand side is extended by match options, positional filters or logical connectives:

```
/library/book[content ftcontains (''biogenetics" ftor
(''biology" ftand ''genetics" ordered distance at most 5 words))
language 'en' with stemming with thesaurus default]
```

Z. Bellahsène et al. (Eds.): XSym 2009, LNCS 5679, pp. 114–128, 2009.

Due to the complexity of the language extension, this paper will focus on its core features. Special attention will be given to the discussion of different execution plans. As full-text requests heavily depend on index structures, the query compiler will try to use a full-text index whenever possible. If this strategy fails, a sequential approach is chosen. A third, hybrid variant takes advantage of the index, but processes all XML nodes sequentially.

While iterative query processing (streaming) adds some overhead to simple database operations, it clearly wins when large intermediate and small final result sets are to be expected. As all XQuery expressions in BaseX are implemented in an iterative manner, the iterative approach was not only maintained for all full-text operators, but even pushed down to the index methods and structures. This way, execution times for small results will not suffer from bulky index results.

Many full-text queries produce large result sets with long textual contents. Since, from the beginning, BaseX supported visual access to data and query results, the graphical frontend was extended to meet the demand of visualizing large text bodies and results in a compact way.

The paper is organized as follows: Section 2 presents the storage and index structures that allow for efficient query evaluation. The sequential, index-based and hybrid execution strategies are discussed in Section 3, and details on iterative query evaluation are given in Section 4. Some performance results in Section 5 analyze execution times of the evaluation variants. Section 6 gives insight into the visual presentation of full-text results; it is concluded by the summary in Section 7.

2 Database Architecture

2.1 Document Storage

While many different XML storage models have been discussed over the last ten years—and none of them has superseded the others—the Pre/Post encoding and its variants have proven to generally yield good performance. It was introduced by Grust [16] and successfully applied by the MonetDB/XQuery implementation [6]. Several variations of this encoding can be used to faithfully represent the XML structure. In MonetDB, for example, XML nodes are mapped to a `pre/size/level` triple. The attributes represent a node identifier, the number of descendant nodes and the depth of a node inside the document tree.

As shown in Figure 1, BaseX stores a `pre/dist/size` combination for each node. The `size` attribute is mainly used to speed up child and descendant traversals, whereas `dist` contains the distance to the parent node, allowing access to the parents and ancestors of a node in constant time. As we will see later, index-based queries benefit greatly from fast access to ancestor nodes. A relative parent encoding (the distance) was favored over an absolute reference as it has shown to be update-invariant, i.e., sub-trees keep their original distance values if they are moved to another place or inserted in a new document.

The main advantage of a flat storage of XML documents is that documents can be sequentially parsed—a property that is particularly useful if many subsequent

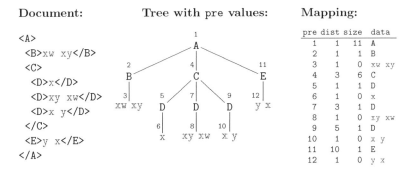

Fig. 1. Document Encoding in BaseX

nodes have to be accessed, which is the case, e.g., for traversals of the `descendant` step. Next to that, the final table contains no variable-sized entries. As tags and attribute names are indexed and texts and attribute values are separately stored, tuples can be stored with a fixed size, and the memory/disk offset of XML nodes can be calculated easily and accessed in constant time [14].

A closer look at the table attributes reveals some specific properties for each node kind (element, attribute, text, etc.):

- the `size` value of text and attribute nodes will always be 0
- the number of distinct tag and attribute names is much smaller than the number of document nodes
- as elements have a limited number of attributes, the `dist` value of attribute nodes is small
- attributes, however, consist of two values (attribute name and value)

Based on these and some other observations, the storage of XML node tuples can be compacted. This compression procedure further speeds up node access by minimizing the tuple sizes.

The presented storage was simplified for the sake of clarity. The actual storage model includes some other data structures, such as a directory to reference the first `pre` values of the disk-based table blocks [14]. This extension is needed to support update operations on the storage. The general access time, however, is not affected by the extension. To get even better performance, the database table can be completely kept in main-memory—a feature which is obviously limited by the amount of available memory.

2.2   Index Structures

The presented storage is extended by a number of index structures. **Name indexes** convert variable-sized tag and attribute names as well as namespaces to fixed-size numeric references. An additional **path summary** maintains information on all distinct location paths in an XML document [3,12]. Both indexes are enriched by statistical data (number of occurrences, minimum and maximum values of attached text nodes/attribute values), which are interpreted by the query optimizer, as shown in Section 3. **Value indexes** reference all text nodes and attribute values of a document. They are used to speed up content-based queries. A classical example for the application of a value index is the combination of a location path filtered by an equality predicate: `/A/C[D = ''x"]`. Query evaluation can be skipped at an early stage, if a value index indicates that a query will yield zero hits. Among others, the attribute index is beneficial to evaluate the XQuery *fn:id()* and *fn:idref()* functions.

To capture the challenges of XQuery Full Text, all text nodes are tokenized, normalized and stored in an additional **full-text index**. The tokenization process is further specified in Section 4.1 of the language specification [1]. Normalization includes the removal of diacritics, a case insensitive representation, optional stemming, etc. A Compressed Trie [2,10] was implemented that, apart from simple token requests, supports flexible operations such as range, wildcard and fuzzy queries. Figure 2 shows a trie struc-

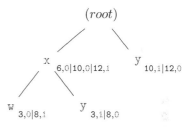

Fig. 2. Compressed Trie: characters with `pre,pos` value pairs

ture (simplified) for the document from Figure 1. Each node contains characters of the indexed `token`, and the `pre,pos` value pairs ($pre_0,pos_0|\ldots|pre_n,pos_n$) identify all occurrences of the token. The `pre` value references the text nodes stored in the database table; the position within the text node is remembered as `pos` value. As the index is built in document order, all stored `pre,pos` values are automatically sorted—a property which comes in handy, as we will see in Section 4.

Whereas many tries are designed to work in main memory, the presented index exclusively operates on flattened and compressed array structures. This way, it can be directly stored to disk, and access time and memory consumption is minimized. As some index requests—such as a *count()* on the number of results—will only access meta data, structural and reference data are stored in separate containers. The structural container contains the indexed token characters, references to child nodes, the number of results, and offsets to the reference container which contains all `pre,pos` pairs. More implementation details can be found in [11].

3   Full-Text Evaluation Strategies

BaseX employs three different evaluation strategies for full-text queries: sequential scanning, index-based processing with path inversion and a hybrid approach. All of them are presented here, along with a decision framework to select the best mode. The following queries are used to illustrate the query evaluation strategies:

Q1: `/A/C[D/text() ftcontains ''x"]`
Q2: `//D[text() ftcontains ''x"]`
Q3: `//*[text() ftcontains ftnot ''x"]`

3.1   Sequential Scanning

Query Q1 consists of child steps and a predicate with an `ftcontains` expression. The corresponding sequential query plan (simplified) is depicted in Figure 3. The evaluation requires a sequential scan of the document. The *LocationPath* expression starts from the root node and traverses all child nodes. Each A element is passed on to the next child step, and the resulting C elements are filtered by the *FTContains* expression. The left-hand *LocationPath* yields all `text()` nodes of D elements, which are checked for the token `''x"`.

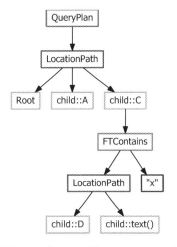

Fig. 3. Query Plan: sequential processing of Query Q1

Obviously, with increasing document size, the sequential scan becomes a bottleneck as all nodes addressed by the query have to be touched at least once.

3.2   Index-Based Processing with Path Inversion

In XML databases, a large variety of index types exists. Content (or value) indexes facilitate direct access to text nodes in a document, and different variants are found in practice:

- Some databases reference results on the **document level**. This is often done if XML is stored in relational database columns. Queries on many small documents can be accelerated by this approach, while there is no benefit for single and large documents.
- Certain **location paths** can be pre-selected for being indexed. While this seems promising at first glance, it often fails when queries are nested or getting more complex. Moreover, users need explicit knowledge about the existing index structure.
- Implementation-defined **XQuery functions** allow for a direct index access. Knowledge on the database internals is needed, and, next to that, a query compiler will not benefit from the indexes, as the user alone decides whether the index is to be used.

To support arbitrary full-text expressions, we chose to index all text nodes by default, regardless of their position in the document structure. As demonstrated in the following, the query optimizer will rewrite and invert location paths and predicates whenever an index access is possible.

In Figure 4, the index-based execution plan of Query Q1 is depicted. In contrast to the sequential scanning mode, which evaluates queries from the document root down to leaf nodes, a bottom-up approach is pursued by first accessing the full-text index and secondly traversing the path back from the leaf nodes to the document root.

First of all, the *FTIndex* operator returns the references of all text nodes containing the token ''x". Next, parent elements D and C are selected. Finally, the ancestor path of the remaining nodes (including the document node) is checked to dismiss results which do not comply with the original query path. Path inversion is possible due to the symmetries of certain XPath axes. Forward-looking, top-down variants have been discussed in detail in [20], and some of them are shown in Table 1. By extending them to multiple location steps, they serve well to dynamically rewrite a large number of location paths.

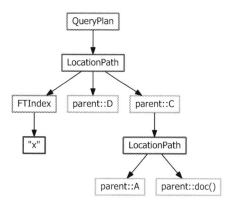

Fig. 4. Query Plan: index-based processing of Query Q1

Table 1. Location paths and their equivalents

Path	Equivalent Path
/descendant-or-self::m/child::n	/descendant::n[parent::m]
/descendant-or-self::m/descendant::n	/descendant::n[ancestor::m]
p[ancestor::m]/self::n	p/self::n[ancestor::m]
p/following::m/descendant::n	p/following::n[ancestor::m]

The second Query Q2 (//D[text() ftcontains ''x"]) introduces a descendant-or-self and child step, which can be merged, in this case, to a single descendant::D step. Queries with descendant steps will be executed more slowly by some query engines, as virtually all document nodes have to be touched and checked for its node kind and tag name. The optimized, index-based execution plan in Figure 5, however, is very compact: as the descendant step in the original query selects all D elements in the document, regardless of their path to the root node, the ancestor and document test can be completely skipped. As the additional ancestor traversal, which has to be evaluated for each single node, takes additional time, this query will be executed even faster than Q1.

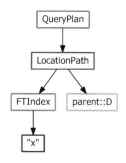

Fig. 5. Query Plan: index-based processing of Query Q2

Value indexes can be used to find out, where a text is found in a document, but not to find places of its absence. If a full-text query contains an ftnot expression, the option to use an index access with consecutive path inversion turns out to be useless.

However, the index can still be of value in a sequential traversal, as the tokenization and normalization of all touched text nodes can take much longer than a simple reference test in a modified `FTNot` operator implementation.

3.3 Hybrid Processing: Sequential Evaluation with Index Usage

Figure 6 shows the resulting query plan for Query Q3 (`//*[text() ftcontains ftnot ''x"]`). It resembles the sequential execution plan—except for the full-text expressions, which are all index-aware. If the *FTIndex* operator is called for the first time, the index is accessed once. *FTIndexNot* checks for each node if it is not part of the index result, and *FTIndexContains* works similar to the conventional *FTContains* operator, but basically avoids tokenizing the current node. If the incoming nodes are guaranteed to be sorted, *FTIndexNot* will operate even faster. As all index references are sorted as well (see Section 2.2), it can completely run in an iterative manner.

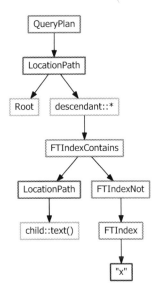

Fig. 6. Query Plan: hybrid processing of Query Q3

3.4 Choosing the Proper Processing Strategy

A two-step model is used by the query compiler for choosing the proper processing strategy. In the first step, it is decided whether it is possible and efficient to use the index, while the second step rewrites the affected operators in a positive first case. The tag/attribute index and path summary are used to perform some basic cost estimations, which influence the decision for or against index access. The number of expected text nodes, their average text length (which influences the time for tokenizing text nodes) and their position in the path summary are considered as well as the number of index results, which can be requested from the full-text index. If a query potentially allows performing several index requests, it can be cheaper to only access the index once and process the other predicates sequentially. Query execution can be completely skipped if the index indicates that a term will yield no results at all.

For the sake of simplicity and to present but the core functionality, we have limited the discussion to the optimization of basic location paths. A slightly more complex query is shown in Figure 7. It contains a *FLWOR* expression, a general comparison and an `ftcontains` expression with an additional `ftand` connective. The query plan illustrates that the available indexes can be applied here as well.

```
let $auction := doc('XMark.xml')
return
    for $p in $auction/site/people/person
  where $p/address/country = 'United States'
    and $p/name ftcontains 'Nikil' ftand
        'Stolovitch' case insensitive
  return $p/emailaddress
```

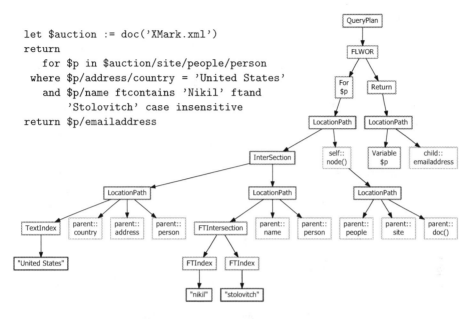

Fig. 7. XQuery with *FLWOR* expression

4 Iterative Evaluation of XQuery Full Text

4.1 Sequential Evaluation

Iterative/pipelined query evaluation is a general database concept [13] which is applied in a number of other XQuery implementations [8,9,18]. In contrast to a conventional, set-based approach, items are processed one-by-one, which guarantees constant memory consumption. The pipeline is only broken by so-called blocking operators that need their complete input, which is the case for sorting, for instance. Iterative evaluation can add some minimal overhead, but it yields particularly good performance when the creation of large intermediate results can be avoided, that are later reduced to a small, final result set.

Although the internal XQuery Full Text data model is complex, as scoring values are calculated and word positions are passed on to evaluate so-called positional filters (such as word order or distances, see [1] for details), all expressions can be evaluated in an iterative manner. The *FTAnd*, *FTOr* and *FTUnaryNot* expressions are implemented similarly to their XQuery counterparts; both processing modes handle one node per iterator step. Consider, e.g., the *FTOr* iterator that merges nodes with equal `pre` values and returns the node with the smallest `pre` value and its corresponding `pos` values. The full-text references of the remaining operands have to be temporarily cached as iterators return data only once. Additionally, the *FTMildNot* operator, which has no XQuery equivalent, has to check whether one occurrence of the first operand is not followed by any other occurrence of the remaining operands.

4.2   Index-Based Full-Text Iterator

As described in Section 2.2, the full-text index references `pre` and `pos` values for each index term. Querying the index means that all references are fetched from disk and returned via an iterator. But in many cases, the entire full-text data is not needed to successfully evaluate a query. Therefore, the iterator concept was pushed down to the index structures. The iterative implementation of the `FTIndex` operator works as follows: After initializing the iterator with the structural data, all data for the first node reference (pre_0) is read, i.e., all `pre,pos` value pairs from pre_0,pos_0 to pre_0, pos_n are processed and returned. In the next iteration, the data stored for the reference pre_1 is read and returned, and so on. This process continues as long as more index results are requested, or all references have been returned.

4.3   Iterator Trees: Processing Non-trivial Index Requests

Iterative index processing is simple and straightforward, as long as single index terms are requested. If the index, returns results for wildcard queries, for instance, the references of several index terms have to be merged and returned. As all index references (i.e., their `pre` value) are sorted by document order, the iterative approach can easily be extended to an arbitrary number of index iterators and a union expression on top of them. Each single index access is managed by an index iterator. It keeps the offset and number of `pre,pos` value pairs stored for an index token.

The following wildcard example is based on the introductory XML document and full-text index shown in Figures 1 and 2. For each index hit, which is recursively matched by the trie algorithm, an index iterator is created. The resulting index tree is evaluated every time an index result is requested. The `pre,pos` value pairs with the smallest `pre` value are merged and returned.

The following example illustrates the presented approach. The full-text query `//*[text() ftcontains ''x.*" with wildcards]` yields all elements with a text node that contains a token starting with the character `''x"`. In our example, three tokens (x, xw, xy) match the wildcard expression. The wildcard algorithm creates an index iterator tree, which is depicted in Figure 8.

Each iterator, which represents one single token, returns results in the known format $pre_0,pos_0|\ldots|pre_n,pos_n$. At the first step, the smallest `pre` values have

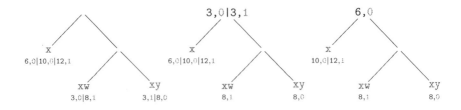

Fig. 8. Index iterator for `//*[text() ftcontains ''x.*" with wildcards]`

to be obtained. Therefore, each node of the iterator tree returns its smallest `pre` value and the corresponding `pos` values. Next, the `pos` references of equal `pre` values are merged. As shown in the figure, the root node now contains the minimum `pre` value 3 and the merged `pos` values 0 and 1. The next step will move `pre` value 6 to the top. After that, the second and third iterator will return their values for `pre` value 8, and the index tree will be reduced to a single iterator, which will return `pre` values 10 and 12.

5   Experimental Analysis

The following tests demonstrate the performance gains by applying indexes to full-text querying. All tests were performed with BaseX 5.6[1]. We used a 2.3 GHz Intel Xeon CPU with 32 GB RAM as hardware and Suse Linux 10.2 and Java 1.5.0.16 as software. Four XMark instances (sized 11 MB, 111 MB, 1 GB and 11 GB) were generated and used as query input.

Table 2. Tested queries

#	Query
Q1	`doc('xmark')//keyword[text() ftcontains 'barrel']`

Q2	`for $mail in doc('xmark')/site/regions/*/item/mailbox/mail`
	`where $mail//text/text() ftcontains 'seeking.*' with wildcards`
	`return $mail/from`

Q3	`for $item in doc('xmark')/site/regions/*/item`
	`where $item//listitem/text/text() ftcontains ftnot 'preventions'`
	`return <result>{ $item/location/text() }</result>`

The three queries in Table 2 are supposed to summarize the discussed query rewritings. Query Q1 contains a simple descendant step and an `ftcontains` expression. Query Q2 uses a number of child steps to address the relevant text nodes, and the full-text expression is extended by a wildcard option. An `ftnot` operator is used in the third query Q3.

All performance results are listed in Table 3 and illustrated in Figure 9. The times represent the average over several runs (5-100 runs, depending on the document size); they include the time for parsing, compiling and evaluating the query as well as printing the result. The boxes show the result sizes in kilobytes.

As expected, all index-based queries yield better results than their sequential equivalents. The index-based version of

Query	11 MB	111 MB	1 GB	11 GB
Q1: Size	0,2	0,7	6	60
Q2: Size	1,5	14	118	1190
Q3: Size	16	165	1656	16602
Q1: Sequential	0.116	1.109	11.03	109.8
Q1: Index	0.001	0.003	0.017	0.128
Q2: Sequential	0.302	2.964	29.39	292.3
Q2: Index	0.006	0.041	0.396	3.831
Q3: Sequential	0.138	1.383	13.43	132.3
Q3: Hybrid	0.074	0.721	7.355	75.15

Table 3. Result size in KB, execution times in seconds

---

[1] Open-source, available at `http://www.basex.org`

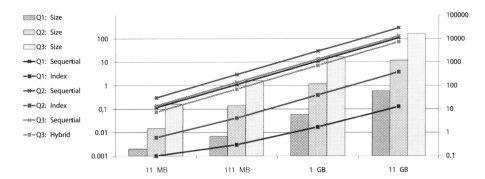

Fig. 9. Performance results. Boxes/right axis: result size in KB, lines/left axis: execution time in seconds.

Q1 is evaluated fastest, as the resulting query plan (which is similar to Figure 5) only contains the index access and a parent step. The scalability is sub-linear, as the index version is about 1000 times faster than the sequential version with the 11 GB input, compared to a factor of 100 for the 11 MB input. Q2 adds some overhead with the wildcard operator, and the larger result size amounts to a virtually linear execution time for both the sequential and the index-based approach. Query Q3 demonstrates the potential of the hybrid query evaluation. As tokenization of text nodes can be avoided, index-supported querying is about twice as fast as pure sequential processing. In spite of the large result size, the hybrid approach is still faster than the pure sequential solution for Query Q1. Documents with larger text nodes (such as, e.g., the Wikipedia XML instances[2]) will yield even better results if text tokenization can be avoided.

As the performance results indicate, there is a clear relationship between the execution times and the data size. As larger XML instances yield larger result sets, it is worth adding that the sequential and hybrid execution is mainly dependent on the size of the input document, whereas the index-based variant exclusively depends on the size of the query result.

6   Visualization of XML and Full-Text Results

Since the first release, BaseX offers a graphical frontend to visually explore content and structure of stored XML data [15,17]. Figure 10 (background) shows a Wikipedia fragment using the Treemap visualization [21]. Each element is drawn as a rectangle and the element tag is printed in the upper left area of this rectangle. The inherent structure of the instance is clearly recognizable: A starting `siteinfo` element containing some meta data, which is followed by several `page` elements each corresponding to a Wikipedia article. The structure of the `page` elements is good to grasp as well: `page` elements contain a `title`, `id` and `revision` element, which again contains elements, for instance the `text`

---

[2] Available at `http://download.wikimedia.org`

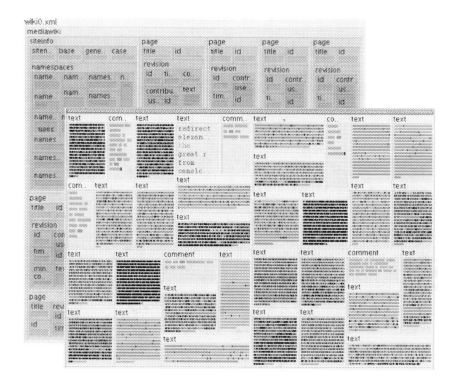

Fig. 10. Treemap visualizations of XML data

element storing the full-text article. The space-filling treemap often allows the viewer to comprehend the complete structure of a document at a glance. By interacting with the treemap, e.g., zooming into a subarea, a higher level of detail can be achieved. As such, an explorative browsing approach may be used to obtain further details about the data instance. Rectangles corresponding to result nodes of a query are highlighted using a contrasting color code.

It is in the nature of full-text queries to often produce large result sets with long textual contents. Our standard text visualizations have shown to be insufficient in terms of compact result presentation and general overview over content and structure. We chose to enhance the treemap visualization by a dynamic abstraction layer using token/sentence thumbnails in combination with full-text tooltips to overcome these deficiencies.

As previously discussed, full-text operators report the `pre` value and the token position `pos` for each search term in a full-text query. Leveraging such information, a visualization can provide a more compact and space-preserving treemap layout by using thumbnail representations for tokens or, at a higher level, sentences. The approach is straightforward. Whenever there is enough space to place the original text into a rectangle, it is displayed as usual. If this is not

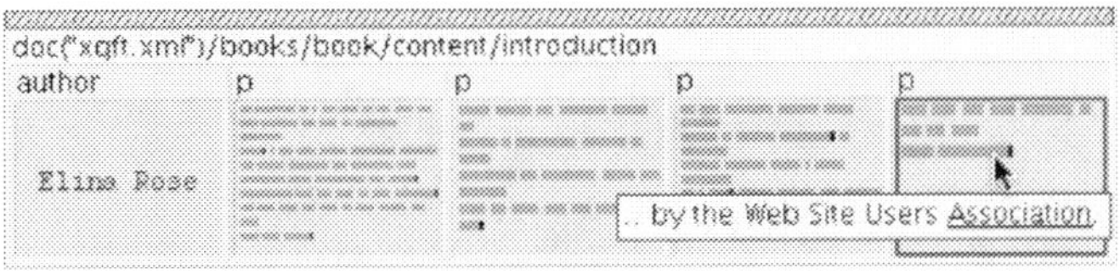

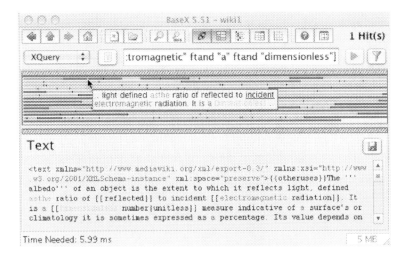

Fig. 11. Full-text thumbnail and tooltip representation

Fig. 12. Split visual result presentation of a full-text query. Above: a sentence based thumbnail representation with highlighted full-text tokens. Below: the textual representation in the original document.

the case, tokens are replaced by thumbnails, following an approach by Kaugars [19]. The length of a thumbnail correlates with the size of the represented text token. Line breaks between tokens are preserved.

Figure 11 illustrates the thumbnail representation. As the textual node of the `author` element fits into the corresponding rectangle, it is displayed in its readable format. The thumbnail representation is used for the text nodes of the `p` elements. The black thumbnail entities denote periods or other sentence terminators. As mentioned, the length of a thumbnail is relative to the length of the represented token, as such the structure of the sentence is preserved. Once the mouse cursor is moved over a thumbnail, the original text is displayed in a tooltip.

Figure 10 (foreground) displays an area of 35 elements in a Wikipedia instance. All occurrences of the term *"the"* are highlighted. The figure demonstrates another abstraction layer (representing a whole sentence by a thumbnail) of the visualization procedure. In one of the treemap rectangles, there is enough space to fit in the textual content (*"redirect alexander the great r from camelcase"*). However, it is yet too narrow to display the tokens *"alexander"* and *"camelcase"* completely, so they are truncated to *"alexan.."* and *"camelc.."*. In the `comment` elements, the token thumbnail representation is chosen. Once more, black rectangles indicate sentence delimiter. For the `text` elements, the sentence-based thumbnail abstraction is chosen. Hereby we can observe two characteristics: For text passages of median length the original sentences are still good to be recognized. The longer text passages are, the darker they appear due to the increasing number of delimiters. The structure of the text, however, is preserved in all abstraction levels.

Using dynamic thumbnail representation for full-text bodies allows space-saving visual representations of large text bodies in a small display area. Combined with tooltips, which additionally display preceding and following text blocks of the selected token, it is possible to sequentially read and browse through the compacted, thumbnailed text, as illustrated in Figure 12.

7 Summary

We presented aspects of the architecture of the XQuery Full Text Recommendation in BaseX, an open-source DBMS developed at U Konstanz. As one of, if not *the*, first complete implementation of all language features, our system provides simple sequential query processing algorithms that allow for pipelined processing of operator sequences as well as (full-text) indexes to speed-up search. In addition, a hybrid query execution strategy is employed whenever pure index-based or sequential processing seems to promise only second-best performance. Substantial query rewrite optimizations have already been incorporated, even though BaseX does not yet involve a full-blown cost-based query optimizer trying to always find the best possible plan.

Our initial performance evaluation proves perfect scalability of both, sequential and index-based execution plans. Actually, we were even able to take advantage of indexes for the evaluation of queries with negated full-text predicates (`Not` expressions). Finally, BaseX's visual querying interface and result display has also been extended for full-text applications, such that matches w.r.t. full-text predicates can be highlighted in query results. Several XML visualizations are available in BaseX, e.g., the treemap that clearly show the document structure together with varying content detail, depending on document or result set size. Using highlights and tooltips or split views, the system gives visual feedback to the user as to where matching part of the XML document have been found.

Future work will include more subtle query optimization and index evaluation strategies as well as additional functionality to cover language-specific full-text features. Also, we plan to extend our visual querying interface and result display with a variety of zoomable representations.

References

1. Amer-Yahia, S., et al.: XQuery and XPath Full Text 1.0. W3C Candidate Recommendation (May 2008), http://www.w3.org/TR/xpath-full-text-10
2. Aoe, J.-I., et al.: An Efficient Implementation of Trie Structures. Software – Practice and Experience 22(9), 695–721 (1992)
3. Barta, A., et al.: Benefits of Path Summaries in an XML Query Optimizer Supporting Multiple Access Methods. In: Proc. of the 31st VLDB Conference, Trondheim, Norway, pp. 133–144 (2005)
4. Bhaskar, A., et al.: Quark: an efficient XQuery full-text implementation. In: Proc. of the ACM SIGMOD Conference, Demo Tracks, Chicago, Illinois, USA, pp. 781–783 (2006)
5. Boag, S., et al.: XQuery 1.0: An XML Query Language. W3C Recommendation (January 2007), http://www.w3.org/TR/xquery
6. Boncz, P.A., et al.: MonetDB/XQuery: a fast XQuery processor powered by a relational engine. In: Proc. of the ACM SIGMOD Conference, Chicago, Illinois, USA, pp. 479–490 (2006)
7. Curtmola, E., et al.: GalaTex: A Conformant Implementation of the XQuery Full-Text Language. In: Proc. of the 2nd XIME Workshop, Baltimore, Maryland, USA (2005)
8. Fischer, P., et al.: MXQuery – a low-footprint, extensible XQuery Engine (2009), http://www.mxquery.org
9. Florescu, D., et al.: The BEA/XQRL Streaming XQuery Processor. In: Proc. of the 29th VLDB Conference, Berlin, Germany, pp. 997–1008 (2003)
10. Fredkin, E.: Trie Memory. J. CACM 3(9), 490–499 (1960)
11. Gath, S.: Processing and Visualizing XML Full-Text Data. Master's thesis, University of Konstanz, Germany (2009)
12. Goldman, R., Widom, J.: DataGuides: Enabling Query Formulation and Optimization in Semistructured Databases. In: Proc. of the 23rd VLDB Conference, Athens, Greece, pp. 436–445 (1997)
13. Graefe, G.: Query Evaluation Techniques for Large Databases. ACM Computing Surveys 25(2), 73–170 (1993)
14. Grün, C., et al.: Pushing XPath Accelerator to its Limits. In: Proc. of the 1st ExpDB Workshop, Chicago, Illinois, USA (2006)
15. Grün, C., et al.: Visually Exploring and Querying XML with BaseX. In: Proc. of the 12th BTW Conference, Demo Tracks, Aachen, Germany, pp. 629–632 (2007)
16. Grust, T.: Accelerating XPath Location Steps. In: Proc. of the ACM SIGMOD Conference, Madison, Wisconsin, USA, pp. 109–120 (2002)
17. Holupirek, A., et al.: BaseX & DeepFS: Joint Storage for Filesystem and Database. In: Proc. of the 12th EDBT Conference, pp. 1108–1111 (2009)
18. Hoschek, W.: Nux – an Open-Source Java toolkit for XML Processing (2006), http://acs.lbl.gov/nux
19. Kaugars, K.J.: A Hierarchical Approach to Detail + Context Views. PhD thesis, New Mexico State University, Las Cruces, NM, USA (1998)
20. Olteanu, D., et al.: XPath: Looking Forward. In: Proc. of the XMLDM Workshop, pp. 109–127. Springer, Heidelberg (2002)
21. Shneiderman, B.: Tree Visualization with Tree-Maps: 2-d Space-Filling Approach. ACM Trans. Graph. 11(1), 92–99 (1992)

Recommending XMLTable Views for XQuery Workloads

Iman Elghandour[1,*], Ashraf Aboulnaga[1], Daniel C. Zilio[2],
and Calisto Zuzarte[2]

[1] University of Waterloo
[2] IBM Toronto Lab

Abstract. Physical structures, for example indexes and materialized views, can improve query execution performance by orders of magnitude. Hence, it is important to choose the right configuration of these physical structures for a given database. In this paper, we discuss the types of materialized views that are suitable for an XML database. We then focus on XMLTable materialized views and present a procedure to recommend them given an XML database and a workload of XQuery queries. We have implemented our XMLTable View Advisor in a prototype version based on IBM[®] DB2[®] V9.7, which supports both relational and XML data, and we experimentally demonstrate the effectiveness of our advisor's recommendations.

1 Introduction

XML is becoming widely adopted as a data storage and representation format. In addition to native XML database systems, most commercial database systems now support an XML column type and have query optimizers that can handle XML data and queries [6,21,22]. Furthermore, these database systems allow creating physical structures such as indexes and materialized views to improve the query execution performance of XML queries. For large databases and complex query workloads, it is challenging to choose the right configuration of physical structures that also have a reasonable disk usage.

Recommending indexes and materialized views as part of the physical database design process has previously been studied extensively in the context of relational databases, and most commercial database systems now include *Design Advisors* that automatically recommend various physical structures [2,23]. The high-level outline of the recommendation process for XML databases is similar to that for relational databases. However, recommending indexes and materialized views for XML databases presents some unique challenges that make the problem more difficult than the relational case, and that lead to the details of the solutions being significantly different.

---

[*] Supported by an IBM PhD Fellowship. Also affiliated with Alexandria University, Alexandria, Egypt.

Z. Bellahsène et al. (Eds.): XSym 2009, LNCS 5679, pp. 129–144, 2009.

There are currently several types of materialized views for XML data. Different proposals have defined different view languages for XML data and have studied matching these views with XML queries. In this paper, we discuss these different approaches and we then focus on one of them, namely XMLTable materialized views. We discuss the advantages of using XMLTable materialized views, which are relational in structure, to improve the performance of XQuery workloads. Next, we present a physical design advisor that recommends XMLTable materialized views for XQuery workloads. We present an experimental study of the the effectiveness of this XMLTable View Advisor.

The main issues that we address when recommending materialized views are: (1) determining the candidate physical structures (materialized views) that would be useful for a query or a workload consisting of a set of queries, (2) expanding the set of candidates by adding new ones that are useful for multiple queries in the workload, and (3) searching the space of possible materialized view configurations for the optimal configuration that provides the maximum benefit to the workload while satisfying disk, schema, and other system constraints. In this paper, we present novel techniques to address each of these challenges. We have implemented our XMLTable View Advisor in a prototype version of DB2 V9.7, which supports both relational and XML databases, and we have used this implementation to verify the efficiency of our proposed advisor and the high quality of the view configurations that it recommends.

The rest of the paper is organized as follows. We present related work in Section 2. Next, we present our contributions, which can be summarized as follows:

- A brief discussion of the existing materialized view languages for XML data (Section 3).
- We propose an end to end solution for an XMLTable View Advisor that recommends relational materialized views that are constructed using the SQL XMLTable function (Section 4). Within our solution for the XMLTable View Advisor we make the following contributions: (1) a technique for enumerating XMLTable views that are useful for an XQuery query (Section 4.2), (2) an algorithm that translates XQuery queries into relational queries that use XMLTable views (Section 4.3), (3) a generalization algorithm that generates new XMLTable views that are useful for multiple queries in the current workload (Section 5), and (4) a search algorithm that extends the heuristic algorithm introduced in [10] to address the interaction between views (Section 6).
- An implementation of the XMLTable View Advisor in a prototype version of DB2 and an experimental study using the TPoX [19] benchmark (Section 7).

2   Related Work

In the past few years, there has been a considerable amount of work on automatic physical design for relational databases [2,23]. Unfortunately, none of these works extend directly to XML databases. The XML Index Advisor proposed in [10] recommends XML indexes for an XML database given a workload of XML queries. Our XML View Advisor expands on the Index Advisor by recommending XMLTable views, which are more complex than the partial XML indexes recommended by the Index Advisor. In this section, we first discuss existing approaches that decide on how to store the data based on its characteristics. Next, we present previous cost based approaches that are used to recommend materialized views for XML databases.

Our approach relies on recommending relational materialized views for XML queries. Relational and XML data reside side by side in current database systems [6]. Query execution cost depends on the storage mode of the data, and so there are situations where it is appropriate to use a relational representation of the data and others where it is appropriate to use an XML representation. A discussion of the factors affecting the choice of using a relational or XML representation to store data is presented in [14,18]. The proposed solution is to find a logical design for a database given the characteristics of the data to be stored in it. However, application access patterns of the data are also important. These access patterns can be exploited to add materialized views to the database to enhance performance [12]. To incorporate both relational and XML data models in the same database system, several hybrid XML-relational architectures are presented in [13].

Another area where relational and XML data coexist is publishing relational data as XML, an area that has been extensively studied in the last few years. In these systems, data is stored in relational stores and published as an XML schema, which requires translating XQuery queries into SQL queries, and translating relational data into XML data that satisfies the published XML schema. Most publishing techniques have one fixed way to translate the relational data into XML based on the XML schema. However, some research projects attempt using a cost based analysis for choosing the best translation [7,9].

In MARS [9], the data is originally stored in relational and XML format, in addition to partial views of the data that are of relational and XML types. In that work, one virtual XML view is published and the incoming queries are translated according to the source that is chosen to answer them. A cost based analysis to choose the best query translation is proposed.

In LegoDB [7], the mapping between XML and relational views of the data is also chosen according to a cost based approach. The application is represented by a workload of queries and data statistics. A subset of the XML schema, called *p-schema*, is used to describe the data. P-schema has the advantage that it can be directly mapped to relational data, and also it is annotated by statistics information. Initially, different candidate p-schemas are enumerated. Then, a greedy heuristic search is used to find the best schema. The cost of a schema is estimated by performing the mapping between the XML data and the relational storage, translating the XML workload according to this mapping, importing the XML statistics into the new relations, and finally, using a relational optimizer to estimate the cost of the workload.

An attempt to partially automate the logical design of a hybrid (Relational-XML) database system is presented in [18]. The input to the proposed *Schema Advisor* is an annotated information model that is considered as a conceptual design for the database. Based on this annotated model, the schema advisor analyzes different storage alternatives and chooses the best of them according to a scoring function. Users of the system can also give their input to the tool to guide the advisor process.

Another cost based approach for automating the logical design of XML databases is proposed in the ULoad project [4]. That work uses the XML Access Modules (XAMs) algebraic formalism to represent the data and its storage structures. ULoad uses a fixed set of designs to choose from, but the users can expand them with their own persistent data structures using the same graphical language. A structural summary of the data is then used to estimate the cost of answering a workload of queries given a configuration of XAMs.

3   Materialized Views for XML Data

Creating views of relational and XML data can take place on either the logical or physical level or both. On the logical design level, data can be XML and be published as relational views [13,17], or data can be relational and be published as XML views [9,17]. Queries are written according to the published schema, so if, for example, the published schema is XML and the data is stored in relational format, we need to (1) translate the XML queries to SQL queries according to the stored schema, and (2) transform the XML data to relational to be stored in the relational store, and vice versa for query answers.

On the physical design level, materialized views of XML data can be in one of the following forms:

1. Views of XML data fragments that are defined by XQuery queries [3,20]. The queries written against the views are also in XQuery. Result containment is checked to decide if a view can answer a query.
2. Views of XML data fragments that are defined by XPath path expressions [5,16]. Queries can be either XPath or full XQuery. In the latter case, indexes containing fragments of the data constitute the XML views.
3. Views of XML data elements and their values that are defined by XPath path expressions and stored in relational tables. XQuery queries are then translated into SQL queries to be executed on these materialized relational views. This approach is close to shredding the XML data into relational tables [7,8]. We adopt this approach in this paper and elaborate on it next.

3.1   XMLTable Views of XML Data

Using relational materialized views for XML data and queries allows us to benefit from the rich and mature infrastructure for these views built into many database systems. Using these views provides a simple and effective way to improve the performance of XML workloads by leveraging existing infrastructure. Building

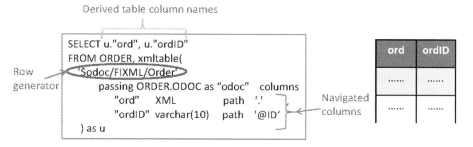

Fig. 1. XMLTable view example

relational views of XML data requires a mechanism that maps between XML elements and their corresponding column names in the relational views. For example, in ROX [13], the XML Wrapper of IBM DB2 [15] is used to do this mapping. The XML Wrapper allows CREATE NICKNAME statements that include nicknames for XPath expressions in the XML document.

A new approach for creating relational views for XML data is to use the XMLTable function [1,21]. XMLTable is a SQL table function that creates a derived table based on XML data. The XMLTable function is applied on a table with an XML-type column. Each row of the table has an XML document in this XML-typed column, and the XMLTable functions maps elements occurring in these XML documents to columns in the derived table generated by the XMLTable function. The parameters of an XMLTable function are: (1) A row generator, which is a path expression. Each element reachable by this path expression corresponds to a tuple in the derived table. (2) Column navigators, which are Xpath navigation patterns. Each column navigator is used to populate a column in the derived table. The row generator specifies the rows in the derived table generated by the XMLTable function, and the column navigators specify the columns of these rows. Figure 1 illustrates an example SQL query with an XMLTable function.

Using the XMLTable function to create relational views of the XML data allows us to benefit from both the mature relational view matching [12] and also XPath view matching [5,16]. The XMLTable is defined in the FROM clause of a SELECT statement which allows two levels of matching of queries with views. The optimizer matches queries that contain XMLTable functions with XMLTable views. Next, XMLTable definitions of the query and view can use XPath matching to find the needed compensation and so to rewrite the query to use the view contents. A discussion of the possible techniques and issues related to matching and rewriting queries to use XMLTable views is presented in [11].

In this paper, our goal is to recommend XMLTable materialized views that benefit a workload of XQuery queries on data that is stored in an XML-typed column of a table. This requires: (1) enumerating XMLTable views for an XQuery query and translating the query to use the views, (2) expanding the set of candidate views, and (3) choosing the best set of views given a disk space budget. We elaborate on these three steps in Sections 4-6, respectively.

4   View Enumeration Process

4.1   Types of XMLTable Views

We employ a cost based analysis to choose the views that would benefit the queries in the workload the most. The high level architecture of the XMLTable View Advisor is as follows. First, we analyze each query in the workload and enumerate its possible XMLTable view candidates. The set of XMLTable views enumerated for all queries in the workload constitutes our basic set of candidate views. Next, we expand the set of candidate views by recommending more general views that can answer more queries in the workload. Then, for each candidate view, we invoke the query optimizer in a special mode to estimate the benefit of the view to the queries in the workload. Finally, we search the space of candidates to find the best configuration of views that has the highest benefit to the workload and fits into the given disk space budget. Our advisor architecture is similar to that of the XML Index Advisor described in [10]. The proposed advisor is based on employing common access patterns of XQuery queries to decide on the views that are useful for them. For example, if a query frequently accesses an element's value in the XML data (an ID for instance), then it is beneficial to extract it as a separate column in the XMLTable view.

The class of XQuery queries that we support includes queries with FOR, LET, WHERE, and RETURN clauses. The RETURN clause can have either a simple or a constructed expression. The general form of a query that we support is as follows:

GQ

```
for $forVar in (ColumnName)/forExpr[forPredicate]
let $letVar := aggFn(letExpr)
where wherePredicate
return returnExpr
```

We use the following query *Q1* on the TPoX [19] benchmark database as a running example:

Q1: For every customer whose age is greater than 50 and has an ID greater than 9000, return her name and the number of accounts she has.

```
for $cust in ("CUSTACC.CADOC")/Customer[@id > 9000]
let $accounts := count($cust/Accounts/Account)
where $cust/age > 50
return
    <print>
        <name>$cust/name</name>
        <accounts_number>$accounts</accounts_number>
    </print>
```

4.2   Enumerating Candidate Views

To enumerate candidate views for an XQuery, we parse the query and break it down into its FOR, LET, WHERE, and RETURN clauses. Then, for each one of these clauses we further break it into its components. We describe next how we handle each clause in the candidate enumeration process (Algorithm 1).

Algorithm 1. enumerateCandidates(*xquery*)

1: **for** *clause* ∈ *xquery* **do**
2:   **if** *clause is forClause* **then**
3:     create a new view *view* and associate it with the variable $*forVar*
4:     set the row generator of *view* to be *forExpr*
5:     **for** *p* ∈ *forPredicate* **do**
6:       add *p* to *view* as a column navigator
7:     **end for**
8:   **else if** *clause is letClause* **then**
9:     create a new view *view* and associate it with the variable $*letVar*
10:     **if** *letExpr* references an existing *refView* **then**
11:       *resolvedLetExpr* ← append the row generator of *refView* and *letExpr*
12:       set the row generator of *view* to be *resolvedLetExpr*
13:       add column "." to *refView* and a backward navigation path to *view* (these columns are used to join the two views *view* and *refView*)
14:     **else**
15:       set the row generator of *view* to be *letExpr*
16:     **end if**
17:     **if** *clause* has *aggFn* **then**
18:       add a SQL GROUP BY clause to *view* with all columns except the expression that appears in the *aggFn*
19:     **end if**
20:   **else if** *clause is whereClause* **then**
21:     **for** *p* ∈ *wherePredicate* **do**
22:       find *refView* that is referenced in *p*
23:       add *p* to *refView* as a column navigator
24:     **end for**
25:   **else if** *clause is returnClause* **then**
26:     **for** *expr* ∈ *returnExpr* **do**
27:       find *refView* that is referenced in *expr*
28:       add *expr* to *refView* as a column navigator
29:     **end for**
30:   **end if**
31: **end for**

FOR Clause. We divide the FOR clause into a variable, a path expression, and its optional predicates. For every FOR clause: (1) we create a new view and assign its row generator to be the path expression extractor in the FOR clause (i.e. the path expression after removing any predicate values from it, *forExpr* in *GQ*), (2) we record the variable name and the created view so we can add any expression that references it to the view as a column, and (3) finally, for every path appearing in a predicate, we create a navigation path and add it to the view. For example, when we parse the FOR clause of *Q1*, we create a new view *V1* that has the row generator /Customer and the column @id:

V1:
```
select u.cx0 from CUSTACC, xmltable(
    '$cadoc/Customer' passing CUSTACC.CADOC as "cadoc"
     columns
            cx0  double path '@id') as u
```

LET Clause. Similar to the FOR clause, we parse the LET clause to find the clause variable (*letVar* in *GQ*) and its binding expression (*letExpr* in *GQ*). In addition, a LET clause might have an optional aggregation function that we only take into account when we rewrite the XQuery to use the view and a binding expression that references a previously bound variable ($cust/Accounts/Account in Q1). For an expression with a reference variable, we look up the expression referenced by this variable (/Customer in this example) and concatenate it with the rest of the expression to form the path expression we use for this clause. We then create a new view with that new path expression as a row generator. We add a column in each of the newly created view and the old one to be used for joining them together in the translated query. The updated version of *V1* and the newly created *V2* will be as follows:

V1:
```
select u.cx0, u.cx1 from CUSTACC, xmltable(
    '$cadoc/Customer' passing CUSTACC.CADOC as "cadoc"
     columns
            cx0  double path '@id',
            cx1  xml path '.') as u
```

V2:
```
select count(u.cy0) as ACc1, u.cy1 from CUSTACC, xmltable(
    '$cadoc/Customer/Accounts/Account' passing CUSTACC.CADOC as "cadoc"
     columns
            cy0  xml path '.',
            cy1  double path 'parent::Accounts/parent::Customer') as u
group by cy1
```

WHERE Clause. For every predicate appearing in a WHERE clause, we handle each predicate expression by finding the view referenced by the variable that appears in this expression, and adding a column to that view to correspond to this navigation. To account for the predicate on age in Q1, view *V1* is now written as follows:

V1:

```
select u.cx0, u.cx1, u.cx2 from CUSTACC, xmltable(
    '$cadoc/Customer' passing CUSTACC.CADOC as "cadoc"
    columns
            cx0   double path '@id',
            cx1   xml path '.',
            cx2   double path 'age') as u
```

RETURN Clause. For all the expressions that appear in the RETURN clause, we find all the variables that reference views and we find the views that they reference. We add a column for each variable to the corresponding view. View *V1* can be updated now to have **name** as a column:

V1:
```
select u.cx0, u.cx1, u.cx2, u.cx3 from CUSTACC, xmltable(
    '$cadoc/Customer' passing CUSTACC.CADOC as "cadoc"
    columns
            cx0   double path '@id',
            cx1   xml path '.',
            cx2   double path 'age',
            cx3   varchar(100) path 'name') as u
```

4.3   Translating XQuery Queries into SQL Queries That Use XMLTable Views

Current XML query optimizers lack the infrastructure to perform the matching of XQuery queries with relational (XMLTable) views. Existing matching algorithms match queries with XMLTable function to XMLTable views [11]. Therefore, we outline in this section a procedure to translate XQuery queries into SQL queries with XMLTable functions. The translation involves using views that are similar to the ones being recommended, and hence we perform the candidate enumeration step described in the previous section to find the best suitable view for a query. Next, we use these recommended views to rewrite the query.

We examine the parsed XQuery, and then construct an SQL query based on this information. We add all the recommended views to the FROM clause of the SQL query. We then use the column names in the views in the SELECT clause and WHERE clause according to the binding of variables and how they appear in the original query. We also add joins between the views that are used to rewrite the query when needed.

For example, we have recommended two views *V1* and *V2* for *Q1* and we can now construct the FROM clause as FROM V1, V2. Next, we examine the return clause and construct the SELECT clause of the rewritten query. If the return value is a simple XPath expression, then the corresponding column name is used, otherwise if an XML fragment is constructed, an XQuery construction is done using the XMLELEMENT function. Finally, we construct the WHERE clause as a conjunction of all the predicates that appear in the XQuery and those that correspond to joins between views. The final rewritten query for Q1 is as follows:

Rewritten Query: RQ1
```
select  XMLELEMENT( NAME  "print" , XMLELEMENT( NAME  "name" ,
     Vv1.cx3) , XMLELEMENT( NAME "accounts_number" , Vv2.ACc1))
from  (..same as V1..) as Vv1, (..same as V2.. ) as Vv2
where ( Vv1.cx2 >  50 )  and  ( Vv1.cx0 >  9000 )
     and ( Vv2.cy1  =  Vv1.cx1 )
```

5   Expanding the Set of Enumerated Views

In the XML Index Advisor [10], we have found that generalizing the index patterns makes them useful for queries not seen in the workload that is used for the recommendation. Similarly, creating views that answer multiple queries in the workload and potential unseen queries can increase the usefulness of our recommendations. Since our proposed view definition encapsulates both XPath expressions and SQL query definitions, generalization can benefit from the index generalization techniques we proposed in [10] and the query merging techniques proposed in [23]. The possible generalization techniques include generalizing the row generator or the column navigator of the view, and merging views. In addition, it is possible to use relational indexes on XMLTable views to increase their benefit. We describe some of the possible query generalization forms that we have explored in this section.

Generalizing Column Navigators to Include Subtrees. Most of the XML-Table views that we recommend in the enumeration phase are a normalization (flattening) of all the values that are being accessed in the workload queries. An alternative approach is to recommend views that store sub-trees of the data as XML columns. A recommended XMLTable view can now have the XPath path expression to reach the data as the row generator and one column with a "." path expression to represent all the subtrees reachable by that row generator. For example, $V3$ (below) is a generalization of $V1$. This approach is useful when the query requires reconstructing the XML tree. This general view requires that the matching infrastructure allows matching multiple columns in the query with one column in the view and is also capable of performing XPath compensation. For example matching view $V3$ with $Vv0$ in query $RQ1$ means matching columns $cx0$, $cx1$, $cx2$ and $cx3$ in $Vv0$ with $cx0$ in $V3$ and requires navigating for @id, age, and name, respectively, in the rewritten query that uses the view. Instead of replacing the columns of a view with a "." column, a less aggressive approach for generalizing column navigators is to consider pairs of views that share the same row generator and consider pairs of columns, one from each view, and generalize these columns together using index generalization algorithms that we propose in [10].

V3:
```
select u.cx0 from CUSTACC, xmltable(
   '$cadoc/Customer' passing CUSTACC.CADOC as "cadoc"
   columns
            cx0  int path '.') as u
```

Merging Views. A common generalization approach used in relational advisors is view merging [23]. For XMLTable views, we merge views that have the same row generator to produce a new view that has the set of column navigators that appear in the merged views after removing duplicates. The goal of this approach is to decrease the disk space required for views by removing duplicate columns from the merged views, while still achieving the same performance.

Indexes on XMLTable Views. One approach to make XMLTable views more useful is to build relational indexes on their columns and hence improve query performance. This is possible since the XMLTable views are defined in the form of SQL statements that produce relations. There can be many possible indexes that can be built on the different columns of an XMLTable view to help the view perform better. In this paper, we use a heuristic approach to select only one index for each view. The chosen index has all the columns of the view that have originally participated in a predicate in the XQuery that caused this view to be recommended. This way we guarantee that these columns have relational values that are used for lookup in the query. The index follows the same order of the columns in the view. For example, the index that we recommend for view *V1* is "`create index index1 on V1(cx0, cx2)`". For every candidate view, we add to the search space another alternative structure which is composed of the view with a relational index over its columns.

6   Searching for the Optimal View Configuration

To recommend a set of XMLTable views (a view configuration) for a workload, we need to search the space of candidate views to find the best set of views that fits into a given disk space budget. We generalize the search algorithms in [10] to be able to search any physical structure (indexes, views, views with indexes on them, etc.). The search problem can be modeled as a 0/1 knapsack problem. The size of the knapsack is the disk space budget specified by the user. Each candidate physical structure – which is an "item" that can be placed in the knapsack – has a *cost*, which is its estimated size, and a *benefit*. We compute the benefit of a physical structure as the difference between the workload cost as estimated by the query optimizer before and after creating this structure.

XMLTable views can interact with each other in ways that reduce their total benefit for a query workload. Our search algorithm takes such interactions into consideration. The main types of interaction affecting the selection of views are: (1) views that can be used together to rewrite a query, and (2) views that are generated by merging other views. These interaction factors are similar to the

---

Algorithm 2. heuristicSearch(*candidates*, *diskConstraint*)

1: sort *candidates* according to their *benefit*(*cand*)/*cand.size* ratio
2: *recommended* ← ∅, *recommended.size* ← 0, *recommended.coverage* ← ∅
3: **while** *recommended.size* < *diskConstraint* **do**
4:      *bestCand* ← pick the next best *cand* in *candidates*
5:      **if** *recommended.coverage* = ∅ **or** *recommended.coverage* ∩ *bestCand.coverage* = φ
        then
6:          *addCandIfSpaceAvl*(*bestCand*,*recommended*)
7:      **else if** *recommended.coverage* ≥ *best.coverage* **then**
8:          *replaceCandIfSpaceAvl* (*bestCand*,*recommended*,*recommended*)
9:      **else**
10:         *overlapConfig* ← *overlapCoverage*(*bestCand*, *recommended*)
11:         *replaceCandIfSpaceAvl* (*bestCand*,*overlapConfig*,*recommended*)
12:     **end if**
13: **end while**
14: **return** *recommended*

---

ones encountered when searching the space of XML indexes, so we use a greedy search algorithm as in [10], but we modify the heuristic rules used in this search to deal with interactions so that they suit the view search problem.

The high level outline of the greedy search algorithm is as follows. First, we estimate the size of each candidate view, and the total benefit of this view for the workload. We then sort the candidate views according to their benefit/size ratio. Finally, we add candidates to the output configuration in sorted order of benefit/size ratio if they agree with the heuristic rules, starting with the highest ratio, and we continue until the available disk space budget is exhausted. In [10] we proposed heuristic rules that are based on *index coverage*. We define the *view coverage* of a view as its view ID as well as the ID of the views that it subsumes. Subsequently, the coverage of a configuration of views is the combination of the view coverage of its constituent views. For example, if *V3* is generated by merging *V1* and *V2*, then the coverage of *V3* is the set of $\{1, 2, 3\}$. We refer to the coverage of a candidate view (*cand*) or a group of views (*config*) as *cand.coverage* and *config.coverage* respectively. We also refer to the size of a candidate view (*cand*) as *cand.size*. Algorithm 2 outlines the the search algorithm. We use the following functions to apply the heuristics and perform the search:

- *benefit*(*config*) returns the estimated benefit of the workload when this configuration of views (or views with relational indexes on them) is created. It is based on calling the query optimizer with and without the views in place and computing the reduction in the optimizer's estimated cost when the views are in place.
- *addCandIfSpaceAvl*(*cand*, *config*) adds the candidate (*cand*) to the configuration (*config*) if the *cand.size* + *config.size* ≤ *diskConstraint*. In addition, if the condition holds, *addCandIfSpaceAvl* updates the *size* and *coverage* of *config*.

- *replaceCandIfSpaceAvl(cand, subConfig, config)* replaces the *subConfig* in *config* with *cand* if the new configuration after performing the replacement *newConfig* has a higher benefit than *config* and the added size is below a threshold β. This is the heuristic that we add to the greedy search to deal with view interactions. The value β is a threshold that specifies how much increase in size we are willing to allow. We have found $\beta = 10\%$ to work well in our experiments. Finally, if the condition holds and there is enough disk space to do the replacement, *replaceCandIfSpaceAvl* updates the *size* and *coverage* of *config*.
- *overlapConfig(config1, config2)* scans a *config2* and returns the minimal *subConfig* configuration that has the view coverage of *config1*.

7   Experiments

7.1   Experimental Setup

Since V9.1, DB2 supports both relational and XML data [6]. We have used an initial prototype version of IBM DB2 V9.7 that was modified to support creating materialized views using the XMLTable function [1]. The client side of the XMLTable View Advisor is implemented in Java 1.6, and communicates with the prototype server via JDBC. We have conducted our experiments on a Dell PowerEdge 2850 server with two Intel Xeon 2.8GHz CPUs (with hyperthreading) and 4GB of memory running SuSE Linux 10. The database is stored on a 146GB 10K RPM SCSI drive.

We used the TPoX [19] benchmark in our experiments. We generate the data using a scale factor of 1GB. We evaluate our advisor on the standard 10 queries that are part of the benchmark specification. We have made minor changes to the workload queries to account for some implementation limitations.

Our XMLTable View Advisor implementation has some limitations due to the existing DB2 prototype infrastructure. These limitations make our advisor unable to recommend views for certain XQuery query types. We can only use SQL data types for columns that appear in the XMLTable functions, since casting XML data into their corresponding relational data types fails in some cases. In addition, columns in XMLTable functions can only be elements; hence, sub-trees reachable by an XPath expression, or linear expressions that select several elements will be concatenated into one large string value. This is not the correct approach when executing XQuery queries. Moreover, our implementation does not support more than two joins per query. We have also left adding support for structured queries, which are XQuery queries with a sub-query in the return clause, for future work. However, these limitations have not prevented us from verifying the usefulness of XMLTable views to answer XQuery queries.

7.2   Effectiveness of the XMLTable View Advisor Recommendations

Figure 2 shows the estimated (based on query optimizer estimates) and actual (based on measured execution time) speedups for the TPoX workload. Speedup

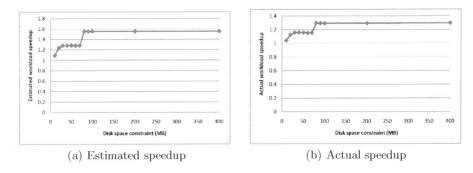

(a) Estimated speedup   (b) Actual speedup

Fig. 2. Workload speedup for the recommended XMLTABLE views

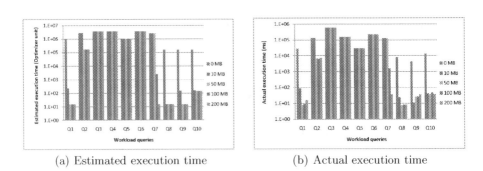

(a) Estimated execution time   (b) Actual execution time

Fig. 3. Query execution time per query for the recommended XMLTABLE views

is defined as the execution time (estimated or actual) of the workload when no XML physical structures are created in the database divided by the execution time of the workload with the view configuration recommended by our advisor is in place. Both figures show that a maximum ratio of 1.6 (for the estimated workload execution speedup) and 1.3 (for the actual workload execution speedup) is achieved when we create the recommended views. Since some queries in the workload did not benefit from views, we also show the estimated and actual execution time of each query in Figure 3. Figures 3(a) and 3(b) show the estimated and actual execution time per query for a configuration with no views and view configurations of different sizes. Queries Q1, Q2, Q7, Q8, Q9, and Q10, which range from simple navigation to join queries, have benefited from the recommended XMLTable views. The actual speedup exceeded 3000 for some queries, for example Q1 and Q7. The configuration which consists of all useful views has a size of 115 MB, which also helped us to achieve an average speedup per query of 639 (the speedup of queries that did not benefit from views is 1). Even for a configuration size of 9.8 MB, the average speedup per query is 134 which proves that XMLTable views can be useful for many query types.

8   Conclusions

In this paper, we have presented an XMLTable View Advisor. This is a new approach for building relational materialized views for XQuery workloads. Our XMLTable View Advisor recommends relational views that are in the form of XMLTable views. These views are useful in pre-navigating to queried values that appear in the data. In addition, XMLTable view matching is based on relational view matching and XPath matching, and hence we benefit from leveraging the already existing infrastructure of many database system query optimizers. We have implemented our advisor in a prototype version of DB2, and our experiments with this implementation show that our advisor can effectively recommend views that result in orders of magnitude performance improvement for some queries.

References

1. XMLTABLE overview (2006),
   http://publib.boulder.ibm.com/infocenter/db2luw/v9/
2. Agrawal, S., Chaudhuri, S., Kollár, L., Marathe, A.P., Narasayya, V.R., Syamala, M.: Database tuning advisor for Microsoft SQL Server 2005. In: VLDB (2004)
3. Arion, A., Benzaken, V., Manolescu, I., Papakonstantinou, Y.: Structured materialized views for XML queries. In: VLDB (2007)
4. Arion, A., Benzaken, V., Manolescu, I., Vijay, R.: ULoad: choosing the right storage for your XML application. In: VLDB (2005)
5. Balmin, A., Özcan, F., Beyer, K., Cochrane, R.J., Pirahesh, H.: A framework for using materialized XPath views in XML query processing. In: VLDB (2004)
6. Beyer, K., et al.: DB2 goes hybrid: Integrating native XML and XQuery with relational data and SQL. IBM Systems Journal 45(2) (2006)
7. Bohannon, P., Freire, J., Haritsa, J.R., Ramanath, M.: LegoDB: Customizing relational storage for XML documents. In: VLDB (2002)
8. Bohannon, P., Freire, J., Roy, P., Siméon, J.: From XML schema to relations: A cost-based approach to XML storage. In: ICDE (2002)
9. Deutsch, A., Tannen, V.: MARS: A system for publishing XML from mixed and redundant storage. In: VLDB (2003)
10. Elghandour, I., Aboulnaga, A., Zilio, D.C., Chiang, F., Balmin, A., Beyer, K., Zuzarte, C.: XML index recommendation with tight optimizer coupling. In: ICDE (2008)
11. Godfrey, P., Gryz, J., Hoppe, A., Ma, W., Zuzarte, C.: Query rewrites with views for XML in DB2. In: ICDE (2009)
12. Halevy, A.Y.: Answering queries using views: A survey. The VLDB Journal 10(4) (2001)
13. Halverson, A., Josifovski, V., Lohman, G.M., Pirahesh, H., Mörschel, M.: ROX: Relational over XML. In: VLDB (2004)
14. Comparing XML and relational storage: A best practices guide. IBM: Storage best practices (2005)
15. Josifovski, V., Massmann, S., Naumann, F.: Super-Fast XML wrapper generation in DB2: A demonstration. In: ICDE (2003)
16. Mandhani, B., Suciu, D.: Query caching and view selection for XML databases. In: VLDB (2005)

17. Manolescu, I., Florescu, D., Kossmann, D.: Answering XML queries on heterogeneous data sources. In: VLDB (2001)
18. Moro, M.M., Lim, L., Chang, Y.-C.: Schema advisor for hybrid relational-XML DBMS. In: SIGMOD (2007)
19. Nicola, M., Kogan, I., Schiefer, B.: An XML transaction processing benchmark. In: SIGMOD (2007), https://sourceforge.net/projects/tpox/
20. Onose, N., Deutsch, A., Papakonstantinou, Y., Curtmola, E.: Rewriting nested XML queries using nested views. In: SIGMOD (2006)
21. Oracle Corp.: Oracle Database 11g Release 1 XML DB Developer's Guide (2007), http://www.oracle.com/pls/db111/
22. Rys, M.: XML and relational database management systems: Inside Microsoft SQL Server 2005. In: SIGMOD (2005)
23. Zilio, D.C., Rao, J., Lightstone, S., Lohman, G.M., Storm, A., Garcia-Arellano, C., Fadden, S.: DB2 design advisor: Integrated automatic physical database design. In: VLDB (2004)

An Encoding of XQuery in Prolog*

Jesús M. Almendros-Jiménez

Dpto. Lenguajes y Computación,
Universidad de Almería
jalmen@ual.es

Abstract. In this paper we describe the implementation of (a subset of) the XQuery language using logic programming (in particular, by means of Prolog). Such implementation has been developed using the Prolog interpreter SWI-Prolog. XML files are handled by means of the XML Library of SWI-Prolog. XPath/XQuery are encoded by means of Prolog rules. Such Prolog rules are executed in order to obtain the answer of the query.

1   Introduction

The *W3C (World Wide Web Consortium)* provides a suitable standard language to express XML document transformations and to query data, the *XQuery* language [14,11,15,10]. *XQuery* is a typed functional language containing *XPath* [13] as a sublanguage. *XPath* supports navigation, selection and extraction of fragments from XML documents. *XQuery* also includes the so-called *flwor* expressions (i.e. `for-let-where-orderby-return` expressions) to construct new XML values and to join multiple documents. *XQuery* has static typed semantics and a formal semantics which is part of the *W3C* standard [11,14].

In this paper we investigate how to implement (a subset of) the *XQuery* language using logic programming (in particular, by means of Prolog). With this aim:

1. XML documents can be handled in Prolog by means of the XML library available in most Prolog interpreters (this is the case, for instance, of *SWI-Prolog* [16] and *CIAO* [9]). Such library allows to load and parse XML files, representing them in Prolog by means of a *Prolog term*.
2. We have to study how to implement XPath and XQuery by means of logic programming. In other words, we have to study how to encode XPath and XQuery queries by means of Prolog rules.
3. Such rules are executed in order to obtain the output of the query. The XML library of Prolog is also used for generating the output.

In previous works we have already studied how to use logic programming for processing XML data. This work continues this research line in the following

---

* This work has been partially supported by the Spanish MICINN under grant TIN2008-06622-C03-03.

Z. Bellahsène et al. (Eds.): XSym 2009, LNCS 5679, pp. 145–155, 2009.

sense. In [4] we have studied how to encode *XPath* by means of rules and how to define a *Magic Set Transformation* [6] in order to execute *XPath* queries by means of *Datalog* following a *Bottom-Up* approach. In [5] we have described how to encode *XPath* by means of rules but in this case, the execution model is *Top-Down*, and therefore *XPath* can be executed by means of Prolog. In [3], we have described how to encode *XQuery* by means of rules, following a *Top-Down* approach, but with the aim to be integrated with the *XPath* encoding studied in [5].

Now, in this paper, we have studied a different approach to the same problem: how to encode *XPath* and *XQuery* by means of rules, but with the aim to integrate the encoding with the XML library available in most Prolog interpreters. Usually, Prolog libraries for XML allow to load a XML document from a file, storing the XML document by means of a *Prolog term* representing the *XML tree*. In our previous works [5,3], XML documents are represented by means of rules and facts. The current proposal uses the encoding of XML documents by means of a Prolog term. The difference of encoding of XML documents has as a consequence that *XPath* and *XQuery* languages have now to be re-encoded for admitting XML documents represented by means of a Prolog term. In order to test our proposal we have developed a prototype which can be downloaded from our Web page http://indalog.ual.es/XQuery.

With respect to existent *XQuery* implementations, our proposal uses as host language a logic language based on rules like Prolog. As far as we know our proposal is the first approach for implementing *XQuery* in logic programming. The existent *XQuery* implementations either use functional programming (with *Objective Caml* as host language) or Relational Database Management Systems (RDBMS).

In the first case, the *Galax* implementation [12] encodes *XQuery* into *Objective Caml*, in particular, encodes *XPath*. Since *XQuery* is a functional language (with some extensions) the main encoding is related with the type system for allowing XML documents and XPath expressions to occur in a functional expression. With this aim an specific type system for handling XML tags, the hierarchical structure of XML, and sequences of XML items is required. In addition, XPath expressions can implemented from this representation. The *XQuery* expressions which do not correspond to pure functional syntax can be also encoded in the host language thanks to the type system.

In our case, SWI-Prolog lacks on a type system, however Prolog is able to handle trees and the hierarchical structure of XML documents by means of Prolog terms. The XML library of SWI-Prolog loads XML documents from a file and represents them by means of a Prolog term of hierarchical structure. XPath is implemented in our approach by traversing the hierarchical structure of the Prolog term. XQuery is implemented by encoding the *flwor* expressions by means of Prolog rules.

In the second case, *XQuery* has been implemented by using a RDBMS. It evolves in most of cases the encoding of XML documents by means of relational tables and the encoding of XPath and XQuery. The most relevant contribution in this research line is *MonetDB/XQuery* [7]. It consists of the *Pathfinder XQuery* compiler [8] on top of the *MonetDB* RDBMS, although *Pathfinder* can be deployed on top of any RDBMS. MonetDB/XQuery encodes the XML tree structure in a relational table following a pre/post order traversal of the tree (with some variant). *XPath* can be implemented from such table-based representation. *XQuery* can be implemented by encoding *flwor* expressions into the *relational algebra*, extended with the so-called *loop-lifted staircase join*.

Our implementation of *XQuery* use as host language Prolog instead of a RDBMS. The advantage of using Prolog is that Prolog can handle the hierarchical structure of a XML document and does not need to encode the tree structure of XML documents. However RDBMS implementations take advantage from optimization techniques for RDBMSs. Since our implementation is based on the encoding into Prolog we are limited in efficiency by the interpreter.

However, our approach has the following advantages. Our aim is the development of a query language for the *Semantic Web*. In this context, XML documents can be handled by means of *XQuery*, however, other kinds of Web documents could be handled in our framework. More concretely, *RDF* and *OWL* documents. A suitable query language for such documents should include reasoning and inference capabilities. Logic programming can be used for Web reasoning. Therefore, a logic programming based implementation of *XQuery* would be easier integrated with rules for Web reasoning. In this line, we have already [1,2] designed extensions of our framework for representing RDF and OWL by means of rules, which can be integrated with our rule based implementation of *XQuery*.

The structure of the paper is as follows. Section 2 will show the representation of XML documents by means of SWI-Prolog; Section 3 will describe the implementation of *XPath* into Prolog; Section 4 will define the translation of *XQuery* expressions into Prolog rules; Section 5 will show the *Eclipse* based tool developed in our framework; finally, Section 6 will conclude and present future work.

2   Loading XML Documents by Means of the Prolog Library

The SWI-Prolog library for loading XML documents stores the XML documents by means of a Prolog term representing a tree. The representation of XML documents is as follows. Each tag is represented as a Prolog term of the form `element(Tag, Attributes, Subelements)` where `Tag` is the name of the XML tag, `Attributes` is a Prolog list containing the attributes, and `Subelements` is a list containing the subelements (i.e. subtrees) of the tag. For instance, let us consider the XML document called "ex.xml" of Figure 1, represented in SWI-Prolog like in the Figure 2.

For loading XML documents in our prototype we can use the predicate `load_xml(+File,-Term)` defined as follows:

```
load_xml(File,Term):-load_structure(File,Term,[dialect(sgml)]).
```

```
<bib>
<book year="1994">
    <title>TCP/IP Illustrated</title>
    <author> <last>Stevens</last><first>W.</first> </author>
    <publisher>Addison-Wesley</publisher>
    <price>65.95</price> </book>
<book year="1992">
    <title>Advanced Programming in the Unix environment</title>
    <author> <last>Stevens</last> <first>W.</first> </author>
    <publisher>Addison-Wesley</publisher>
    <price>65.95</price> </book>
<book year="2000">
    <title>Data on the Web</title>
    <author> <last>Abiteboul</last> <first>Serge</first> </author>
    <author> <last>Buneman</last> <first>Peter</first> </author>
    <author> <last>Suciu</last> <first>Dan</first> </author>
    <publisher>Morgan Kaufmann Publishers</publisher>
    <price>39.95</price> </book>
<book year="1999">
    <title>The Economics of Technology and Content for Digital TV</title>
    <editor> <last>Gerbarg</last> <first>Darcy</first>
        <affiliation>CITI</affiliation> </editor>
    <publisher>Kluwer Academic Publishers</publisher>
    <price>129.95</price>
</book> </bib>
```

Fig. 1. XML document

```
[element(bib, [],
[element(book, [year=1994],
[element(title, [], [TCP/IP Illustrated]),
element(author, [], [element(last, [], [Stevens]), element(first, [], [W.])]),
element(publisher, [], [Addison-Wesley]),
element(price, [], [65.95]) ]),
element(book, [year=1992],
[element(title, [], [Advanced Programming in the Unix environment]),
element(author, [], [element(last, [], [Stevens]), element(first, [], [W.])]),
element(publisher, [], [Addison-Wesley]),
element(price, [], [65.95]) ]),
element(book, [year=2000],
... ])]
```

Fig. 2. Representation of XML in SWI-Prolog

where `load_structure(+File,-Term,+Options)` is the SWI-Prolog predicate of the XML library for loading *SGML* documents. Similarly, we have implemented a predicate called `write_xml(+File,+Term)` for writing Prolog terms representing a XML document into a file.

3   Implementing XPath by Means of Prolog

Now, we will present how *XPath* can be implemented by means of Prolog. We restrict ourselves to *XPath* expressions of the form /*tag*$_1$ ... /*tag*$_n$ (/*text*()). More complex *XPath* queries [13] can be expressed in *XQuery*, and therefore this restriction does not reduce the expressivity power of our proposal. In Prolog,

XPath expressions will be represented by means of lists of the form $[tag_1, \ldots, tag_n,$ $(text)]$ in such a way that we have a predicate load_xpath(+XPath,-ListXPath) to transform Path expressions into the Prolog representation.

Now, the XPath language can be implemented in Prolog by means of a predicate xpath(+ListXPath,+Tree,-Subtrees), where ListXPath is the Prolog representation of an XPath expression, Tree is an input XML document and Subtrees is a list of subtrees of the input document. Basically, the xpath predicate traverses the Prolog tree representing a XML document and extracts in a Prolog list the subtrees occurring in the given path. The predicate includes the following rules, distinguishing cases in the form of the input document and the XPath expression[1]:

```
xpath([text],[Tree|Trees],[Tree|Trees2]):-atomic(Tree),!,xpath([text],Trees,Trees2).
xpath([text],[_|Trees],Trees2):-!,xpath([text],Trees,Trees2).
xpath([Tag],[element(Tag,Attr,SubTrees)|Trees],[element(Tag,Attr,SubTrees)|Trees2]):-!,
                            xpath([Tag],Trees,Trees2).
xpath([Tag],[_|Trees],Trees2):-!,xpath([Tag],Trees,Trees2).
```

For instance, the following goal extracts the subtrees in the path *'bib/book/title'* from the document *'ex.xml'*, and writes them into the file *'output.xml'*:

```
?-load_xml('ex.xml',Term), load_xpath('bib/book/title',LXPath),
            xpath(LXPath,Term,OutputTerm), write_xml('output.xml',OutputTerm).
```

The previous goal generates the following sequence of items:

```
<title> TCP/IP Illustrated< /title>
<title> Advanced Programming in the Unix environment< /title>
<title> Data on the Web< /title>
<title> The Economics of Technology and Content for Digital TV< /title>
```

4   Implementing XQuery by Means of Prolog

Now, we will show how to encode *XQuery* in Prolog using the representation of XML documents and the previous XPath implementation. We will focus on a subset of *XQuery*, called *XQuery core language*, whose grammar can be defined as follows.

Core XQuery

$xquery := dxpath \mid < tag >' \{'xquery, \ldots, xquery'\}' < /tag > \mid flwr.$
$dxpath := doc(Doc)$ '/' $xpath$.
$flwr := \textbf{for } \$var \textbf{ in } vxpath \textit{ [\textbf{where } constraint] } \textbf{return } xqvar \mid$
$\qquad \textbf{let } \$var := vxpath \textit{ [\textbf{where } constraint] } \textbf{return } xqvar.$
$xqvar := vxpath \mid < tag >' \{'xqvar, \ldots, xqvar'\}' < /tag > \mid flwr.$
$vxpath := \$var \mid \$var$ '/' $xpath \mid dxpath$.
$xpath := text() \mid tag \mid xpath$ '/' $tag.$ $Op := <= \mid >= \mid < \mid > \mid =.$
$constraint := vxpath \; Op \; value \mid vxpath \; Op \; vxpath \mid constraint \; Op \; constraint$

---

[1] From now on, we will show the main rules of each predicate, a full version can be downloaded from http::/indalog.ual.es/XQuery

In the previous definition *value* is an *string, integer*, etc, *Doc* is a document name, and *Op* is a boolean operator. The previous subset of the language allows to express the following query:

```
<result>
for $Book in doc('ex.xml')/bib/book return
let $Year := $Book/book/@year
where $Year < 1995 return
<mybook> { $Year $Book/book/title } </mybook>
< /result>
```

Such query requests the year and title of books published before than 1995. It represents the **result** as a sequence of XML items whose tag is **mybook**. The answer of the query is:

<result>
<mybook>1994<title>TCP/IP Illustrated< /title>< /mybook>
<mybook>1992<title>Advanced Programming in the Unix environment</title>< /mybook>
< /result>

In order to encode *XQuery* into Prolog rules we have to take into account the following elements:

- The main element of the encoding is a predicate called **xquery/1** such that **xquery** returns the XML tree representing the result of a query. For instance, the previous query is executed by means of the goal ?- **xquery(Tree)** and the Prolog answer is **Tree=[element(result, [],[element(mybook,[], [1994, ...].**
- In order to define **xquery/1**, we have defined a predicate **xquery/2** of the form **xquery(-Tree,+Number)** where **Tree** is a Prolog term representing a XML tree and **Number** is an identifier of the XML tree **Tree**. In order to build the hierarchical structure of the output tree, each subtree is computed by means to a call **xquery(-Subtree,+Number)**, assuming subtrees are *numbered by levels*. Therefore, the structure of the **xquery** rules is as follows:

xquery([element(tag,[],Subtrees)],Number):-xquery(Subtree,Number+1),
xquery(SubtreeList,Number+2),
combine([Subtree,SubtreeList],Combination),
member(Subtrees,Combination).

whenever the element *tag* has as subtrees in the output document the elements *Subtree* and *SubtreeList*. The root of the output tree is numbered as 1. Therefore **xquery/1** is defined as **xquery(Tree):-xquery(Tree,1)**. When a subtree is a sequence of elements the call to **xquery** is combined with a call to the Prolog predicate **findall**. For instance, *findall(Eachelement, xquery(Eachelement,Number+1),Subtree)*.

– The predicates `xquery` might call to the predicates `flwr(-Tree,+Number)` which compute a *flwr* expression. `Tree` is the output of such expression, and `Number` is the identifier of the `Tree`. In general, the structure of `flwr` predicates is:

---

flwr(Tree,Number):-for_exp(Tree,path(xqueryterm,xpath)).
flwr(Tree,Number):-let_exp(Tree,path(xqueryterm,xpath)).

The predicates `flwr` call to the predicates `for_exp` (and `let_exp`), whenever the *flwr* expression is a `for` expression and `let` expression, respectively. *xqueryterm* is (i) either a variable of the form '*\$X*' or (ii) a document name of the form *doc(docname)*. *xpath* is an *XPath* expression. The meaning of `flwr(Tree,Number)` is that `Tree` (whose number is `Number`) is a Prolog term whose value is the result of evaluating the "pseudo-expression":

(i) "*for Tree in xqueryterm/xpath*" (and "*let Tree := xqueryterm/xpath*"), whenever *xqueryterm* is a document name.

(ii) "*for Tree in doc(docname)/pathtodoc*" (and "*let Tree := doc(docname) / pathtodoc*"), whenever *xqueryterm* has the form '*\$X*'. Where the document name associated to '*\$X*' is *doc(docname)*, and the path from '*\$X*' to the document name is *pathtodoc*.

Let us remark that in our core language, each variable has an associated document name, that is, each variable is used for traversing a given input document. In addition, in case (ii), analysing the *XQuery* expression, a path from the root of the document to the variable can be rebuilt.

– *XQuery* expressions of the form "*doc(Doc)/xpath*" are represented in Prolog as *xpath('doc(docname)',xpath)* and *XQuery* boolean conditions "*\$X/xpath1 Op \$Y/xpath2*" are represented in Prolog as *varpath('\$X',xpath1) Op varpath('\$Y',xpath2)*.

– Our encoding makes a previous program transformation in which *XQuery* expressions including a **return** expression involving *XPath* expressions, are transformed into the so-called *XPath-free return XQuery* expressions, which are equivalent. It will be explained in the following example.

For instance, the previous query can be encoded as follows. Firstly, our encoding transforms the query into an equivalent *XPath-free return* XQuery expression:

```
<result>
for $Book in doc('ex.xml')/bib/book return
let $Year := $Book/book/@year
where $Year < 1995 return
let $Book1 := $Book/book/title return
<mybook> { $Year $Book1 } </mybook>
< /result>
```

where a new variable `$Book1` is introduced by means of a **let** expression in such a way that now, the **return** expression does not include *XPath* expressions. Now, the encoding is as follows:

```
(1)  xquery([element(result, [], A)], 1) :- xquery(A, 2).
(2)  xquery(B, 2) :- findall(A, xquery([A], 3), B).
(3)  xquery([element(mybook, [], A)], 3) :- xquery(A, 6).
(4)  xquery(E, 6) :- findall(A, xquery([A], 7), C), findall(B, xquery(B, 8), D),
       combine([C, D], F), member(E, F).
(5)  xquery([A], 7) :- flwr(B, 7), member(A, B).
(6)  xquery([A], 8) :- flwr(B, 8), member(A, B).
(7)  flwr(A, 7) :- for_exp(A, path('$Year', '')).
(8)  flwr(A, 8) :- for_exp(A, path('$Book1', '')).
(9)  for_exp(B, path(A, C)) :- atomic(A), is_var(A, _), !, for_var(r, B, path(A, C)).
(10) for_var(r, A, path('$Year', C)) :- xquery(B, 5), for_exp(A, path(B, C)).
(11) for_var(r, A, path('$Book1', C)) :- xquery(B, 9), for_exp(A, path(B, C)).
(12) xquery([A], 5) :- flwr(B, 5), member(A, B).
(13) xquery([A], 9) :- flwr(B, 9), member(A, B).
(14) flwr(A, 5) :- let_exp(A, path('$Book', 'book/year')).
(15) flwr(A, 9) :- let_exp(A, path('$Book', 'book/title')).
(16) let_exp(B, path(A, C)) :- atomic(A), is_var(A, _), !, let_var(r, B, path(A, C)).
(17) let_var(r, B, path('$Book', C)) :- xquery(A, 4),
       where_exp(r, '$Book', A, [varpath('$Book', 'book/year')<'1995']),
       let_exp(B, path(A, C)).
(18) xquery([A], 4) :- flwr(B, 4), member(A, B).
(19) flwr(A, 4) :- for_exp(A, path('doc('ex.xml')', 'bib/book')).
(20) for_exp(C, path(A, D)) :- atomic(A), string_to_term(A, doc(B)), !,
       execute_term(B, C, D).
(21) execute_term(A, E, B) :- load_xml(A, D), load_xpath(B, C), xpath(C, D, E).
```

The previous encoding can be summarized as follows:

– The rule (**1**) is the root of the encoding. It defines the Prolog tree element (result, [], A) as the root of the output XML document, where the subtrees are computed in A by means of the rule (**2**).

– The rule (**2**) defines the subtrees of element(result, [], A). They are included in a Prolog list of trees and they are computed by means of the rule (**3**). The rule (**3**) computes the elements enclosed in the tag mybook.

– The rule (**3**) computes the elements element(mybook, [], A). The subelements A of element(mybook, [], A) are couples of elements (representing $Year, $Book1) which are computed by means of the rule (**4**).

– The values of $Year and $Book1 have to be computed by means of a *flwr* expression. The rules (**5**) and (**6**) call to the flwr predicate, defined by means of rules (**7**) and (**8**). Following case (ii), rules (**7**) and (**8**) represent *"for A in doc('ex.xm')/bib/book/@year"* and *"for A in doc('ex.xm')/bib/book/title"*, respectively, given that: the document name associated to *$Year* and *$Book1* is *doc('ex.xml')*; the path from *$Year* to *doc('ex.xml')* is *'/bib/book/@year'*, and the path from *$Book1* to *doc('ex.xml')* is *'/bib/book/title'*.

– With the aim to obtain the previous behaviour the for_exp predicate (in rules (**7**) and (**8**)) calls by means of the rules (**9**), (**10**), (**11**), (**12**) and (**13**) to let_exp predicate in rules (**14**) and (**15**). Rules (**14**) and (**15**) compute the value of the **let** expressions in which $Year and $Book1 are involved.

– The rules (**14**) and (**15**) call to the rule (**16**), which in its turn calls to the rule (**17**). The rule (**17**) calls to rules (**18**) and (**19**) in order to compute the main **for** expression. Following (i) of previous description, the rule (**19**)

Fig. 3. Eclipse-XQuery-SWI-Prolog Tool

represents *"for A in doc('ex.xml')/bib/book"* which is the main *flwr* expression of the query. Moreover, the rule **(17)** checks the boolean condition of the *XQuery* expression, that is, *"$Year < 1995"*.

– Finally, the rules **(20)** and **(21)** compute the main **for** expression by means of the xpath predicate, defined in previous section.

5   A Tool for XQuery

Finally, we would like to show the main features of the tool we have designed for *XQuery*. The tool has been developed taken as basis the *Eclipse* tool, in which we have installed the PDT plugin for SWI-Prolog (available from http://sewiki.iai.uni-bonn.de/research/pdt/users/start). In addition, the *Eclipse* distribution provides support for graphical representation of XML documents. In Figure 3 we can see an snapshot of the tool.

In order to execute *XQuery* expressions in the tool, we have to proceed as follows:

- The *XQuery* expression is loaded into the *Eclipse* tool.
- Input XML documents can be loaded into *Eclipse* tool in order to be visualized.
- A very simple configuration file (a Prolog program) called "query.pl" has to be modified.
- The configuration file is executed by means of the SWI-Prolog plugin.
- Finally, output XML documents can be loaded into *Eclipse* tool in order to be visualized.

6   Conclusions and Future Work

In this paper, we have studied how to encode *XQuery* expressions into Prolog. It allows us to evaluate *XQuery* expressions against XML documents using logic rules. We have developed a prototype available in `http://indalog.ual.es/XQuery`. The distribution includes a package of examples of XQuery expressions which has been tested with our prototype. As future work we would like to extend our prototype for reasoning with RDF/OWL in *XQuery*. The theoretical background of such extension has been studied in [1,2].

References

1. Almendros-Jiménez, J.M.: An RDF Query Language based on Logic Programming. Electronic Notes in Theoretical Computer Science 200(3) (2008)
2. Almendros-Jiménez, J.M.: Ontology Querying and Reasoning with XQuery. In: Proceedings of the PLAN-X 2009: Programming Language Techniques for XML (2009), `http://db.ucsd.edu/planx2009/papers.html`
3. Almendros-Jiménez, J.M., Becerra-Terón, A., Enciso-Baños, F.J.: Integrating XQuery and Logic Programming. In: INAP 2007. LNCS (LNAI), vol. 5437, pp. 117–135. Springer, Heidelberg (2009)
4. Almendros-Jiménez, J.M., Becerra-Terón, A., Enciso-Baños, F.J.: Magic sets for the XPath language. Journal of Universal Computer Science 12(11), 1651–1678 (2006)
5. Almendros-Jiménez, J.M., Becerra-Terón, A., Enciso-Baños, F.J.: Querying XML documents in logic programming. Journal of Theory and Practice of Logic Programming 8(3), 323–361 (2008)
6. Beeri, C., Ramakrishnan, R.: On the Power of Magic. Journal of Logic Programming, JLP 10(3,4), 255–299 (1991)
7. Boncz, P., Grust, T., van Keulen, M., Manegold, S., Rittinger, J., Teubner, J.: MonetDB/XQuery: a fast XQuery processor powered by a relational engine. In: Proceedings of the 2006 ACM SIGMOD international conference on Management of data, pp. 479–490. ACM, New York (2006)
8. Boncz, P.A., Grust, T., van Keulen, M., Manegold, S., Rittinger, J., Teubner, J.: Pathfinder: XQuery - The Relational Way. In: Proc. of the International Conference on Very Large Databases, pp. 1322–1325. ACM Press, New York (2005)

9. Cabeza, D., Hermenegildo, M.: Distributed WWW Programming using (Ciao-) Prolog and the PiLLoW Library. Theory and Practice of Logic Programming 1(3), 251–282 (2001)
10. Chamberlin, D.: XQuery: An XML Query Language. IBM Systems Journal 41(4), 597–615 (2002)
11. Chamberlin, D., Draper, D., Fernández, M., Kay, M., Robie, J., Rys, M., Simeon, J., Tivy, J., Wadler, P.: XQuery from the Experts. Addison Wesley, Boston (2004)
12. Marian, A., Simeon, J.: Projecting XML Documents. In: Proc. of International Conference on Very Large Databases, Burlington, USA, pp. 213–224. Morgan Kaufmann, San Francisco (2003)
13. W3C. XML Path Language (XPath) 2.0. Technical report (2007), www.w3.org
14. W3C. XML Query Working Group and XSL Working Group, XQuery 1.0: An XML Query Language. Technical report (2007), www.w3.org
15. Wadler, P.: XQuery: A Typed Functional Language for Querying XML. In: Jeuring, J., Jones, S.L.P. (eds.) AFP 2002. LNCS, vol. 2638, pp. 188–212. Springer, Heidelberg (2003)
16. Wielemaker, J.: SWI-Prolog SGML/XML Parser, Version 2.0.5. Technical report, Human Computer-Studies (HCS), University of Amsterdam (March 2005)

Universal XForms for Dynamic XQuery Generation

Susan Malaika[1] and Keith Wells[2]

[1] IBM, 19 Skyline Drive,
Hawthorne, NY 10532
[2] IBM, 3039 Cornwallis Road,
Research Triangle Park, NC 27709
{malaika,wellsk}@us.ibm.com

Abstract. This demonstration illustrates how a variety of queries can be built dynamically by examining XML stored in databases through general purpose XForms. The forms are called the Universal XForms for XQuery, and abbreviated to the Universal XForms. The Universal XForms help users to construct XQueries through the provision of prompts, and do not require prior knowledge of the structure of the data to be queried. Sample XML documents from more than twenty industry formats constitute the base for illustrating the building of queries in the demonstration.

Keywords: XML, XQuery, XForms, Database.

1 Introduction

XForms [1] is a W3C XML standard for presenting and collecting XML data. XForms provides the human interaction and entry for business-critical data as well as mechanisms for ensuring the correctness of the data with constraints and relevancy. XForms can be used to visualize and provide input data, and as illustrated in this article, to help users to define queries.

XQuery [2] is another W3C XML standard designed to query and manipulate collections of XML data. For example, queries can be specified to search for all authors in a set of XML instances stored in an XML database such as DB2 pureXML [3] or eXist [4].

Increasingly XML industry formats [5], defined by industry consortia, are used for information exchange because of the useful characteristics of XML, e.g., the ability to define an extensible schema that can be combined with other schemas, the ease with which XML can be processed without needing to consult its schema. Gradually, the exchanged XML is also being stored, e.g., for audit or analysis.

When XML is stored in a database it becomes easy to give access to the stored data through Web Services in a general way. The simplicity and ease arise from the fact that the format of the stored data typically matches what the services need to expose and consume. In other words, what is being stored is being exchanged and matches external world entities such as insurance policies, product descriptions, and tax forms. Thus, less server side customization is needed to support the services. The Universal Services for pureXML [6] provide a simple set of general purpose services to access

Z. Bellahsène et al. (Eds.): XSym 2009, LNCS 5679, pp. 156–164, 2009.

and manipulate stored XML. One of the operations in the Universal Services is an XQuery service.

The demonstration [7] described in this article illustrates how an XForms document [8] can provide assistance in dynamically building XQueries to access collections of stored XML. The XForms document uses the XQuery service in the Universal Services to examine the stored XML and to provide suitable XQuery prompts. The general purpose XForms document does not require prior knowledge of the stored XML documents' structure in order to generate syntactically correct XQueries. The XForms then uses the same XQuery service that was the basis for dynamically building the XQuery to execute the generated XQuery.

The demonstration illustrates how an XML end-to-end architecture that incorporates an XML database, along with declarative markup simplifies tools and application development.

2 The Demonstration Architecture and Principles

The demonstration is built on a database that stores collections of well-formed XML in columns in tables. The stored XML data is exposed in a variety of ways including Universal Services (a fixed set of Web and RESTful Services). The demonstration is built from general-purpose XForms and XQuery requests that are transmitted through the XQuery service that is part of the Universal Services.

The demonstration has been built, rendered and tested with the Firefox XForms Extension [9], but it could easily be adapted for other XForms processors such as Ubiquity XForms [10]. The Ubiquity-XForms open source project is an AJAX/JavaScript implementation of the W3C XForms 1.1 specification. With Ubiquity, XForms such as this could be rendered on any JavaScript-enabled browser; anywhere, anytime, for universal access.

The Universal XForms for Dynamic XQuery Generation are part of a larger demonstration based on industry formats. You can build your own general purpose XML end-to-end environment by following the steps described in Build an Application in a Day [11].

One benefit of using XForms and XQuery with Universal Services over Java, is that declarative markup is edited for building queries, which is then downloaded and executed on the client. There is no need to modify any server side code in order to make available a new or customized version of the Universal XForms for XQuery. All that is required is that the modified Universal XForms be available for download. XML declarative languages are used throughout the demonstrations, without converting the XML into intermediate Java objects. Note also that XForms make it possible to add and remove controls because users made certain selections, e.g., the addition of prompts for where clauses. Only the client (browser XForms) runtime is required to achieve this flexibility.

The demonstration aims to produce syntactically correct XQuery requests that are relevant to the underlying data, enabling authors to experiment and try out queries. However, there is a second variant of the general purpose XForms labeled "debug". In the debug version, the SOAP Envelope for both the Universal XQuery Service Request and Response are shown in an xforms:textarea. An author developing an XQuery expression can view the SOAP Request and Response messages of the Universal XQuery Service.

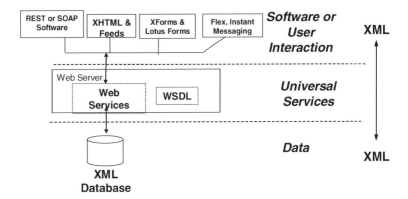

Fig. 1. Diagram of the Demonstration Architecture

3 Stepping through the Demonstration

In this demonstration, XQueries are built dynamically in steps by examining the contents of an XML collection and generating prompts to create the pieces for the next portion of the XQuery expression. As a user makes choices in building an XQuery expression, generated XQueries are executed with the results manipulated via XForms to prompt a user for more information or to determine the next step in building an XQuery expression, until finally, an XQuery expression is generated to reveal the results from user's choices in previous steps.

This SOAP Envelope is used through-out this XQuery Demonstration:

```
<?xml version="1.0" encoding="UTF-8"?>
<soapenv:Envelope soapenv:encodingStyle=
"http://schemas.xmlsoap.org/soap/encoding/" xmlns:soapenv=
"http://schemas.xmlsoap.org/soap/envelope/">
    <soapenv:Header/>
    <soapenv:Body>
        <runxqueryXML xmlns:xsd= "http://www.w3.org/2001/XMLSchema"
xmlns:xsi= "http://www.w3.org/2001/XMLSchema-instance">
            <query/>
        </runxqueryXML>
    </soapenv:Body>
</soapenv:Envelope>
```

In following sections, the <query/> element may be modified in the above SOAP Envelope to invoke an XQuery Universal service.

3.1 Select the XML Collection to Query

What Happens. Because the demonstration is built on an XQuery enabled database system that contains collections of XML in tables, the first step is to select the database table and column name to identify the collection in the XQuery.

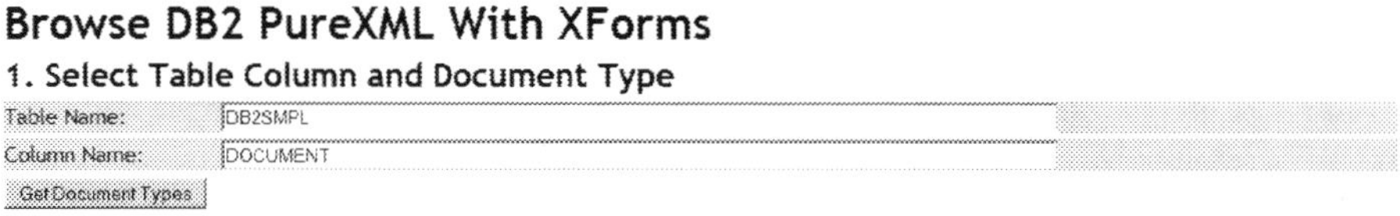

Browse DB2 PureXML With XForms

1. Select Table Column and Document Type

Table Name:	DB2SMPL
Column Name:	DOCUMENT

Get Document Types

Fig. 2. Browse DB2 pureXML With XForms

How It Works. In the XForms document, there is an <xforms:bind> element which binds a calculate attribute to the <query/> element in the XQuery SOAP Envelope.

```
<xforms:bind node-
set="instance('instance_docTypes_Envelope')/soapenv:Body/*[local-
name() = 'runxqueryXML']/*[local-name() = 'query']"
  calculate="concat(
    concat(
      concat(
        concat( in-
stance('queryText_docTypes_instance')/queryParts/prefix,
instance('tableInformation_instance')),
        '.'),
    instance('columnInformation_instance')),
    instance('queryText_docTypes_instance')/queryParts/postfix)"
id="docTypes_queryString_Bind"/>
```

The calculate attribute in this xforms:bind element is building the XQuery expression by concatenating different fragments together using a combination of XPath and XForms functions. The input fields for the Table Name and the Column Name are bound to a specific instance data node. The data node changes when the user modifies the input fields. In pseudo-code, the query is built with basic string concatenation.

```
query = prefix data + table + "." + columnName;
```

And finally the query element is transformed into an XQuery expression.

```
query = "fn:distinct-values( for $x in db2-fn:xmlcolumn(" + table +
"." + columnName + ")/* return <DocumentType> { fn:local-name($x) }
</DocumentType> )"
```

The following is an example of the pre- and post-processing strings used in the above example.

```
<xforms:instance id="queryText_docTypes_instance">
  <queryInformation>
    <query/>
    <queryParts>
      <prefix> fn:distinct-values( for $x in db2-fn:xmlcolumn(&quot;
</prefix>
      <postfix> &quot;)/* return &lt;DocumentType&gt; { fn:local-
name($x) } &lt;/DocumentType&gt; )</postfix>
    </queryParts>
  </queryInformation>
</xforms:instance>
```

When the "Get Document Types" button is selected, an XForms submission invokes the Universal XQuery Service with the query defined from above. Once the resulting web service call returns, the data will be parsed and used in the next step.

3.2   Select the Document Types to Be Queried

What Happens. The column can contain a varied set of well-formed XML documents. The general purpose XQuery forms enables the identification of the document subset with a particular root element to query. The document selected should be typical of the documents that are to be queried, in that it contains representative elements and attributes, to ensure appropriate prompts appear.

As a result of the previous step, all of the root elements queried from XML documents in pureXML for the combination of the Table Name and the Column Name were returned, added to XForms instance data, and are now available to the xforms:select1 control. You should see "product", "customerinfo", "PurchaseOrder" and "a" in this list.

Browse DB2 PureXML With XForms

1. Select Table Column and Document Type

Fig. 3. Select the Document to be Queried

How It Works. In this step, a user selects one of the XML root elements from the list and selects the Get Available Fields button. This starts an XForms submission process to execute the following XQuery:

```
query = let $x :=
db2-fn:xmlcolumn(" + Table Name + '.' + Column Name + ")/*
[local-name()=' + Root Name + '] return $x[1]
```

If "PurchaseOrder" was selected from the xforms:select1 list, the result is the following:

```
<result>
    <PurchaseOrder
        xmlns="http://posample.org"
            PoNum="5000" OrderDate="2006-02-18" Status="Unshipped">
        <item>
            <partid>100-100-01</partid>
            <name>Snow Shovel, Basic 22 inch</name>
            <quantity>3</quantity>
            <price>9.99</price>
        </item>
        <item>
            <partid>100-103-01</partid>
            <name>Snow Shovel, Super Deluxe 26 inch</name>
            <quantity>5</quantity>
            <price>49.99</price>
        </item>
    </PurchaseOrder>
</result>
```

3.3   Select the Elements and Attributes to Be Queried

What Happens. The document collection will typically contain many XML elements and attributes. These are displayed with a prompt to create XQuery predicates. In the following screenshot, you can see that both attributes and child elements of the root node "product" selected from the XQuery in the previous step are presented in the "Select a Field" xforms:select1 list. At this point, a user can choose which attribute or child element of the "product" root node to add to the XQuery in one of the following steps.

For example, a user selects the "name" attribute and then selects the Add Field Entry button. Immediately, the user should see the appearance of additional "Select a Field" prompts, allowing the user to select another Field for the XQuery being built. In this step the user is determining which attribute or child element to query in all of the DB2SMPL XML documents in pureXML database. Once the XQuery is built and executed, these selections are displayed as tabular columns.

Browse DB2 PureXML With XForms

1. Select Table Column and Document Type

Table Name:	DB2SMPL
Column Name:	DOCUMENT
Get Document Types	
Available Document Types:	product
Get Available Fields	

2. Build the Query

Select a Field:	name
Add Field Entry	pid
	name
Order By:	details
Add Where Clause?	price
	weight

3. Execute the Query

| Execute Query |

4. View the Results

Fig. 4. Build the Query

How It Works. The attributes and child elements of the root node "product" were gathered from the results of the XQuery in the previous step, and dynamically added into an xforms:instance fragment. An xforms:repeat and associated xforms:insert and xforms:delete allows for the addition and deletion of the "Select a Field" row.

3.4   Select an "Order by" Clause

What Happens. A selection list is displayed with elements and attributes in the documents that can be used to order the results.

How It Works. The attributes and child elements collected in a previous step are presented in an xforms:select1 list, if a user decides to select one of these choices, an "order by" statement with be concatenated to the XQuery string based on the user's selection.

3.5 Select Optional Where Clauses

What Happens. When an "Add Where Clause?" is selected, extra input fields appear to allow a user to select an element or attribute from the existing documents, together with predicates.

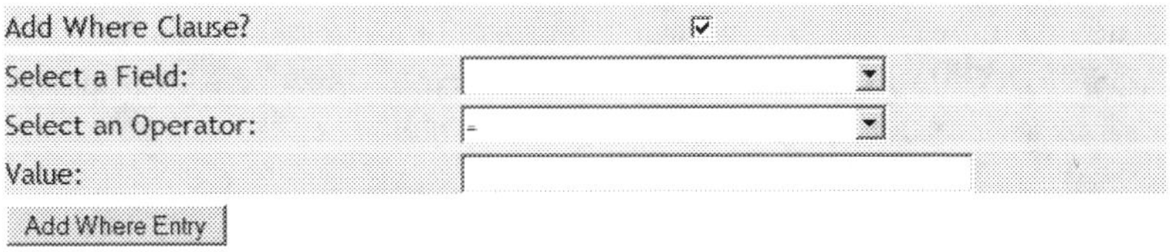

Add Where Clause?	☑
Select a Field:	▼
Select an Operator:	= ▼
Value:	
Add Where Entry	

Fig. 5. Use Where Clauses

How It Works. If a user decides to add a where clause to the XQuery being built, then the "Add Where Clause?" should be selected. Once a user checks this box, additional prompts are displayed to guide the user through the process of building a where clause for the XQuery expression. In this case, the user can select an attribute or child element collected in a previous step, select the discrete operator for the where clause, and add a string value for the where clause comparison. For example, a user may decide to build a where clause to search ·for all price elements which are less than $20.00. In this case, a where clause like the following will be added to the XQuery expression:

```
where price < 20.00
```

3.6 Execute the Query

What Happens. When the "Execute Query" button is selected, the generated XQuery will run. An example of an XQuery could be:

```
for $x in db2-fn:xmlcolumn( "DB2SMPL.DOCUMENT" )/*:PurchaseOrder
where $x/*:item/*:price < '20.00' and $x/*:item/*:price < '20.00'
order by $x/*:item/*:price
return <row> <entry nodeName="OrderDate"> { fn:data($x/@OrderDate) }
</entry> </row>
```

How It Works. A user would "select" the "Execute Query" button when satisfied with the selections up to this point. The "Execute Query" button triggers an xforms: submission invocation of the Universal XQuery service with the generated query dynamically created by the user.

3.7 View the Results

After the XQuery is executed and the response returned, the columns for each field defined during the "Select a Field" step will be displayed as table columns.

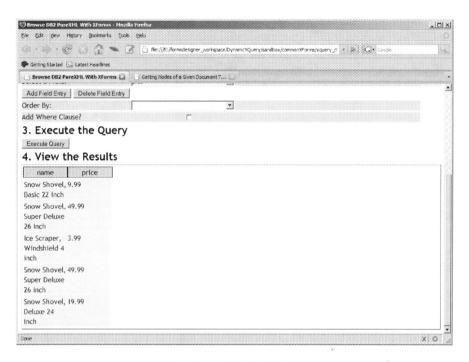

Fig. 6. View the Results

3.8 Possible Enhancements

There are a number of possible extensions to the Universal XForms. For example:

- Sample a few document to base XQuery generation on, or provide an option for users to define the sample a few documents to query.
- Select a few documents that contain a greater variety of elements and attributes to form the basis of the XQuery prompts.
- Enable a more complete XQuery generation, e.g., to include joins, construction, and aggregation. Some queries can become quite complex and become difficult to manually track. In cases like this, using an XQuery builder to set XQuery variables and template XML for the results could save time and frustration.
- Produce various layouts for query outputs, and provide mechanisms to handle large outputs. Perhaps XQuery results could be tailored or customized for a specific layout using HTML, SVG, XForms or some other rendering technique.

4 Summary

This article describes a demonstration that uses XForms to generate queries dynamically that access XML instances stored in a database, without prior information on the structure of the documents. The demonstration illustrates how an XML end-to-end architecture, in conjunction with XML standard technologies, makes it easier to build software solutions.

References

1. XForms at W3C, http://www.w3.org/MarkUp/Forms
2. XQuery at W3C, http://www.w3.org/TR/xquery
3. DB2 pureXML, http://www.ibm.com/software/data/db2/xml/
4. eXist, http://exist.sourceforge.net/
5. Get started with Industry Formats and Services with pureXML,
   http://www.ibm.com/developerworks/db2/library/
   techarticle/dm-0705malaika
6. Universal Services for pureXML using Data Web Services,
   http://www.ibm.com/developerworks/db2/library/
   techarticle/dm-0805malaika
7. pureXML Industry Formats and Services Demonstration,
   http://www.alphaworks.ibm.com/tech/purexml
8. XForms and DB2 pureXML,
   http://www.ibm.com/developerworks/db2/library/
   techarticle/dm-0805malaika2
9. Mozilla XForms Extension,
   https://addons.mozilla.org/en-US/firefox/addon/824
10. Ubiquity XForms, http://code.google.com/p/ubiquity-xforms
11. Build an Application in a Day,
    http://www.ibm.com/developerworks/data/library/
    techarticle/dm-0812malaika

From Entity Relationship to XML Schema: A Graph-Theoretic Approach

Massimo Franceschet, Donatella Gubiani, Angelo Montanari, and Carla Piazza

Department of Mathematics and Computer Science, University of Udine
Via delle Scienze 206, 33100 Udine, Italy

Abstract. We propose a mapping from the Enhanced Entity Relationship conceptual model to the W3C XML Schema Language with the following properties: information and integrity constraints are preserved, no redundance is introduced, different hierarchical views of the conceptual information are available, the resulting XML structure is highly connected, and the design is reversible. We investigate two different ways to nest the XML structure: a maximum connectivity nesting, that minimizes the number of schema constraints used in the mapping of the conceptual schema reducing the validation overhead, and a maximum depth nesting, that keeps low the number of (expensive) join operations that are necessary to reconstruct the information at query time using the mapped schema. We propose a graph-theoretic linear-time algorithm to find a maximum connectivity nesting and show that finding a maximum depth nesting is NP-complete. We complement our investigation with an implementation of the devised translation and we embed the implemented module in a software framework for the conceptual and logical design of spatio-temporal databases.

1 Introduction

The inventors of XML intended to create a document format for web pages and other narrative documents to be read by people. Despite these intentions, the most common applications of XML today involve the storage and exchange of data for use by computer applications. An XML database is a data persistence software that allows one to store data in XML format. Two major classes of XML database exist: *XML-enabled databases*, which map XML data to a traditional database (such as a relational database), and *native XML databases*, which define a logical model for an XML document and stores and retrieves documents according to that method.

The design of a database follows a consolidated methodology comprising conceptual, logical, and physical modeling of the data. This paper is a contribution toward the development of design methodologies and tools for native XML databases. We adopt, in the spirit of [1], the well-understood Entity Relationship model, extended with specialization (ER for short), as the conceptual model for native XML databases. Specialization is particularly relevant in the design of semi-structured data. Moreover, we choose W3C XML Schema Language

Z. Bellahsène et al. (Eds.): XSym 2009, LNCS 5679, pp. 165–179, 2009.

(XML Schema, hereinafter), as the schema language for XML. The major alternative is Document Type Definition (DTD), but DTD is strictly less expressive than XML Schema; in particular, it lacks expressive means to specify integrity constraints, which are fundamental in database design. The contributions of this paper are precisely the following:

1. we propose a mapping from ER to XML Schema with the following properties: information and integrity constraints of the ER model are preserved, no redundancy is introduced, different hierarchical views of the conceptual information are permitted, the resulting structure is highly connected, and the design is reversible;

2. we give a graph-theoretic interpretation of the structure nesting problem, that is, the problem of finding the best way to nest the elements corresponding to entities and relationships of the ER schema. We propose a linear-time algorithm to find a maximum connectivity nesting forest, that is, a nesting forest with the highest number of edges, or, equivalently, with the lowest number of trees. This is the forest that minimizes the number of schema constraints used in the mapping of the conceptual schema and hence that reduces the validation overhead to the minimum. Moreover, we show that the problem of finding a maximum depth nesting forest, that is, a nesting forest with the largest value for the summation of node depths, is NP-complete. Such a forest minimizes the number of (expensive) join operations that are necessary to reconstruct the information at query time using the mapped schema and thus reduces the query evaluation time;

3. we implement the devised mapping and embed it into ChronoGeoGraph [2], a software framework for the conceptual and logical design of spatio-temporal XML and relational databases.

The outline of the papers follows. Section 2 contains the basic mapping from ER to XML Schema. The structure nesting problem is investigated in Section 3. Section 4 summarizes related work and proposes future research directions.

2   The Basic Mapping

This section describes the target schema language and provides the mapping of the basic elements of an ER diagram into the target schema language.

2.1   The Target Schema Language

The XML data model is *hierarchical* and *semistructured*. XML elements may be simple elements containing character data or they may nest other child elements, generating a hierarchical tree-like structure. Moreover, elements of the same type may have different structures, e.g., some child elements may be absent or repeated an arbitrary number of times. By contrast, the relational data model is *flat* and *structured*: table attributes must be atomic and tables have a rigid schema. We will deeply exploit the hierarchical and semistructured nature of the XML data model in the proposed encoding of the ER conceptual model.

Representing XML Schema using its own syntax requires substantial space and the reader (and sometimes the developer as well) gets lost in the implementation details. We embed ER schemas into a more succinct target schema language for XML documents (*XSL* for short) whose expressive power lies between DTD and XML Schema. XSL allows sequences and choices of elements as in DTD. XSL extends DTD with the following three constructs:

- *occurrence constraints.* These constrain the minimum and maximum number of occurrences of an item. The minimum constraint is a natural number and the maximum constraint is a natural number or the constant N denoting a finite unbounded natural number. The notation is `item[x,y]`, where x is the minimum constraint, y is the maximum constraint, and item is a single element, a sequence, or a choice. When both x and y are equal to 1, the occurrence constraint may be omitted;
- *key constraints.* If A is an element and KA is a child element or attribute of A, then the notation `KEY(A.KA)` means that KA is a key for element A. Keys composed of more than one attribute are allowed;
- *foreign key constraints.* If A is an element with key KA, B is an element, and FKA is a child element or attribute of B, then `KEYREF(B.FKA --> A.KA)` means that FKA is a foreign key of B referring to the key KA of A.

Recall that DTD allows only the specification of [0,1] occurrence constraints (denoted by ?), [0,N] occurrence constraints (denoted by *), and [1,N] occurrence constraints (denoted by +). Moreover, DTD offers a limited key/foreign key mechanism by using ID-type and IDREF-type attributes. However, the ID/IDREF mechanism is too simple for our goals, for instance it is not possible to restrict the scope of uniqueness for ID attributes to a fragment of the entire document. Also, only individual attributes can be used as keys.

The mapping of XSL into W3C XML Schema is achieved as follows: sequence and choice constructs correspond to *sequence* and *choice* schema elements; occurrence constraints are implemented with *minOccurs* and *maxOccurs* schema attributes; key and foreign key constraints are captured by *key* and *keyref* schema elements, respectively. A full example of XSL definition in given in Figure 2.

2.2   Mapping ER Elements

An ER schema contains entities and relationships between entities [3]. Both can have attributes, which can be either simple, composed, or multi-valued. Some entities are weak and are identified by proprietary entities through identifying relationships. Moreover, some entities may be specialized into more specific entities. Specializations may be partial or total, disjoint or overlapping. Relationships may involve two or more entities; each entity participates in a relationship with a minimum and a maximum participation constraint. Integrity constraints associated with an ER schema comprise multi-valued attribute occurrence constraints, relationship participation and cardinality ratio constraints, specialization constraints (sub-entity inclusion, partial/total, disjoint/overlapping constraints), as well as key and foreign key constraints. We refer to integrity constraints successfully represented in the target schema language as *internal constraints*, whereas *external constraints* are those constraints that cannot be captured in the target schema language due to lack of expressive power and must be validated using an additional schema validator.

The mapping we propose has the following properties:

- it preserves all the information and as much as possible of integrity constraints of the original conceptual schema. An extension to the standard XML Schema validator has been implemented in order to capture the constraints that are missed in the translation due to lack of expressiveness of the target schema language;
- it does not include redundancy in the mapped schema: the original conceptual information is represented only once in the logical XML design;
- it allows different hierarchical views of the conceptual information; this permits to adapt the structure of the logical schema taking into consideration the typical (most frequent) transactions of the database management system;
- it achieves maximum connectivity in the nesting structure used to embed the elements of the conceptual design. As we will show in Section 3, this amounts to minimize the number of schema constraints used in the mapped schema and hence the validation overhead;
- it allows to reverse the design: from the logical XML schema it is possible to go back to the conceptual ER model.

Entities. Each entity is mapped to an element with the same name. Entity attributes are mapped to child elements. The encoding of composed and multi-valued attributes takes advantage of the flexibility of the XML data model: composed attributes are translated by embedding the sub-attribute elements into the composed attribute element; multi-valued attributes are encoded using suitable occurrence constraints. An example is given in Figure 2. As opposed to the relational mapping, no restructuring of the schema is necessary.

Relationships. Each binary relationship has two (left and right) participation (or cardinality) constraints of the form (x, y), where x is a natural number and stands for the minimum participation constraint, and y is a positive natural number or the special character N that represents a finite and unbounded number and stands for the maximum participation (or cardinality ratio) constraint. Typically, x is either 0 or 1, and y is either 1 or N. Hence, we have $2^4 = 16$ typical cases.

Let us consider two entities A, with key KA, and B, with key KB, and a binary relation R between A and B with left participation constraint (x_1, y_1) and right participation constraint (x_2, y_2). We denote such a case with the notation $A \overset{(x_1,y_1)}{\longleftrightarrow} R \overset{(x_2,y_2)}{\longleftrightarrow} B$. The encodings for all the typical cases are given in the following:

1. $A \xleftrightarrow{(0,1)} R \xleftrightarrow{(0,1)} B$. We have the two possible mappings shown below:

```
A(KA, R[0,1])                    B(KB, R[0,1])
 R(KB)                            R(KA)
B(KB)                            A(KA)
KEY(A.KA), KEY(B.KB), KEY(R.KB)  KEY(B.KB), KEY(A.KA), KEY(R.KA)
KEYREF(R.KB --> B.KB)            KEYREF(R.KA --> A.KA)
```

The two mappings are equivalent in terms of number of used constraints. Notice that the constraint KEY(R.KB) is used to capture the right maximum participation constraint in the left mapping: it forces the elements KB of R to be unique, that is, each B element is assigned to at most one A element by the relation R. Similarly for the constraint KEY(R.KA) in the right solution.

2. $A \xleftrightarrow{(0,1)} R \xleftrightarrow{(0,N)} B$. The preferred mapping is shown on the left below. The solution on the right uses an extra constraint to capture the left maximum cardinality.

```
A(KA, R[0,1])                    B(KB, R[0,N])
 R(KB)                            R(KA)
B(KB)                            A(KA)
KEY(A.KA), KEY(B.KB)            KEY(B.KB), KEY(A.KA), KEY(R.KA)
KEYREF(R.KB --> B.KB)            KEYREF(R.KA --> A.KA)
```

3. $A \xleftrightarrow{(0,1)} R \xleftrightarrow{(1,1)} B$. The suggested view is the left one shown below. The element A is fully embedded into element B; hence no foreign key constraint is necessary and the right minimum cardinality holds. The right maximum cardinality is captured by KEY(B.KB). The solution given on the right uses an additional key constraint for the left maximum cardinality as well as an extra foreign key constraint; moreover, it looses the chance to nest the resulting structure.

```
A(KA, R[0,1])                    B(KB, R[1,1])
 R(B)                             R(KA)
   B(KB)                         A(KA)
KEY(A.KA), KEY(B.KB)            KEY(B.KB), KEY(A.KA), KEY(R.KA)
                                 KEYREF(R.KA --> A.KA)
```

4. $A \xleftrightarrow{(0,1)} R \xleftrightarrow{(1,N)} B$. The suggested mapping is given on the left below. The constraint KEY(R.KA) is used to capture the left maximum cardinality. The opposite solution misses the encoding of the right minimum cardinality, which must be dealt with as an external constraint (added with clause CHECK).

```
B(KB, R[1,N])                    A(KA, R[0,1])
 R(KA)                            R(KB)
A(KA)                            B(KB)
KEY(B.KB), KEY(A.KA), KEY(R.KA)  KEY(A.KA), KEY(B.KB)
KEYREF(R.KA --> A.KA)            KEYREF(R.KB --> B.KB)
                                 CHECK("Right min card")
```

5. $A \overset{(0,N)}{\longleftrightarrow} R \overset{(0,N)}{\longleftrightarrow} B$. We have two symmetrical views using the same number of constraints:

```
A(KA, R[0,N])                    B(KB, R[0,N])
 R(KB)                            R(KA)
B(KB)                            A(KA)
KEY(A.KA), KEY(B.KB)            KEY(B.KB), KEY(A.KA)
KEYREF(R.KB --> B.KB)          KEYREF(R.KA --> A.KA)
```

6. $A \overset{(0,N)}{\longleftrightarrow} R \overset{(1,1)}{\longleftrightarrow} B$. The best solution is the left one below that uses the full nesting of elements. The opposite embedding, on the right, spends an extra keyref constraint and does not achieve element nesting.

```
A(KA, R[0,N])                    B(KB, R[1,1])
 R(B)                             R(KA)
   B(KB)                         A(KA)
KEY(A.KA), KEY(B.KB)            KEY(B.KB), KEY(A.KA)
                                KEYREF(R.KA --> A.KA)
```

7. $A \overset{(0,N)}{\longleftrightarrow} R \overset{(1,N)}{\longleftrightarrow} B$. The mapping on the left is the one we propose. The opposite embedding looses the right minimum participation constraint.

```
B(KB, R[1,N])                    A(KA, R[0,N])
 R(KA)                            R(KB)
A(KA)                           B(KB)
KEY(B.KB), KEY(A.KA)            KEY(A.KA), KEY(B.KB)
KEYREF(R.KA --> A.KA)          KEYREF(R.KB --> B.KB)
                                CHECK("Right min card")
```

8. $A \overset{(1,1)}{\longleftrightarrow} R \overset{(1,1)}{\longleftrightarrow} B$. Two symmetrical views are possible:

```
A(KA, R[1,1])                    B(KB, R[1,1])
 R(B)                             R(A)
   B(KB)                           A(KA)
KEY(A.KA), KEY(B.KB)            KEY(B.KB), KEY(A.KA)
```

9. $A \overset{(1,1)}{\longleftrightarrow} R \overset{(1,N)}{\longleftrightarrow} B$. The preferred mapping is the one given on the left below. The opposite embedding fails to capture the right minimum cardinality.

```
B(KB, R[1,N])                    A(KA, R[1,1])
 R(A)                             R(KB)
   A(KA)                         B(KB)
KEY(B.KB), KEY(A.KA)            KEY(A.KA), KEY(B.KB)
                                KEYREF(R.KB --> B.KB)
                                CHECK("Right min card")
```

10. $A \overset{(1,N)}{\longleftrightarrow} R \overset{(1,N)}{\longleftrightarrow} B$. Two symmetrical mapping are possible:

```
A(KA, R[1,N])                    B(KB, R[1,N])
 R(KB)                           R(KA)
B(KB)                            A(KA)
KEY(A.KA), KEY(B.KB)            KEY(B.KB), KEY(A.KA)
KEYREF(R.KB --> B.KB)          KEYREF(R.KA --> A.KA)
CHECK("Right min card")         CHECK("Left min card")
```

Notice the use of an external constraint in both solutions to check the minimum participation constraint: this is the only case in the mapping of relationships in which we have to resort to external constraints in the preferred mapping. One may be tempted to add, in the left case, the foreign key KEYREF(B.KB --> R.KB) to check the missing constraint. The foreign key would force each B instance to appear under an A instance and hence each B instance would be associated with at least one A instance. Unfortunately, such a foreign key is allowed in XML Schema only if R.KB is a key, which is not possible since a B instance may be associated with more than one A instance and hence there may exist repeated B instances under A. An alternative bi-directional solution is the one that pairs the two described mappings. Such a solution captures all integrity constraints specified at conceptual level. It imposes, however, the verification of an additional *inverse relationship constraint*, namely, if an instance x of A is inside an instance y of B, then y must be inside x in the inverse relationship. Such a constraint is not expressible in XML Schema.

The other six cases are the inverse of some of the above-described solutions. For instance, $A \overset{(0,N)}{\longleftrightarrow} R \overset{(0,1)}{\longleftrightarrow} B$ is the inverse of case 2. We have implemented a similar strategy to map relationships of higher arity and relationships with non-typical participation constraints, e.g., the constraint (2,10).

It is interesting to notice that, thanks to its hierarchical nature, the XML logical model allows to capture more constraints specified at conceptual level than the relational logical model. Indeed, for *all* relationships with one participation constraint equal to (1,N), the minimum participation constraint is lost when mapping the ER model into the relational model; furthermore, some constraints in specialization are also missed [3].

Weak entities and identifying relationships. A weak entity always participates in the identifying relationship with participation constraint equal to (1,1). Hence, depending on the form of the second participation constraint, one of the cases discussed above applies. The key of the weak entity is obtained by composing its partial key with the key of the owner entity and the owner key in the weak entity must match the corresponding key in the owner entity. For instance, suppose we have $A \overset{(0,N)}{\longleftrightarrow} R \overset{(1,1)}{\longleftrightarrow} B$ where B is weak and owned by A. The translation is:

```
A(KA, R[0,N])
 R(B)
  B(KB, KA)
KEY(A.KA), KEY(B.KB, B.KA)
CHECK(B.KA = A.KA)
```

The external constraint `CHECK(B.KA = A.KA)` cannot be avoided. Indeed, suppose we remove the owner key KA from the weak entity B. We need to set a key composed by the pair KA of A and KB of B that now lie at different nesting levels. If we point the selector of the key schema element at the level of entity A, then the field pointing to KB is invalid since it selects more than one node; on the other hand, if we point the selector at the level of entity B, then the field referring to KA is also not valid, since it must use the parent axis to ascend the tree, but such an axis is not admitted in the XPath subset supported by W3C XML Schema[1].

Specialization. The mapping fully exploits the hierarchical nature of the XML data model. Let us consider an entity A with key KA that specializes in two entities B with attributes attB and C with attributes attC. If the specialization is partial-overlapping, then the mapping is as follows:

```
A(KA, B[0,1], C[0,1])
  B(attB)
  C(attC)
KEY(A.KA)
```

Both B and C elements are embedded inside A element. Neither key nor foreign key constraints are necessary. The partial-overlapping constraint is captured by using the occurrence specifiers: an A element may contain any subset of {B, C}. If the specialization is total-overlapping, the regular expression in the first clause must be replaced with `(B, C[0,1]) | C`: any *non-empty* subset of {B, C} is permitted. In case of partial-disjoint specialization the regular expression becomes `(B | C)[0,1]`: either B or C or none of them are included. Finally, a total-disjoint specialization in encoded with the regular expression `(B | C)`: either B or C is present.

The generalization to the case of n sub-entities is immediate in all cases except the total-overlapping case. Let $a_1, \ldots, a_n$ be the sub-entities of a total-overlapping specialization. We indicate with $\rho(a_1, \ldots, a_n)$ the regular expression allowing all non-empty subsets of sub-entities. Such an expression can be recursively defined as follows:

$$\rho(a_1, \ldots, a_n) = \begin{cases} a_1 & \text{if } n = 1 \\ (a_1, a_2[0,1], \ldots, a_n[0,1]) \mid \rho(a_2, \ldots, a_n) & \text{if } n > 1 \end{cases}$$

---

[1] See the official W3C XML Schema Language specification at the W3C site `http://www.w3.org/TR/xmlschema-1/#coss-identity-constraint`. The authors of [4] seem to have repeatedly missed this point.

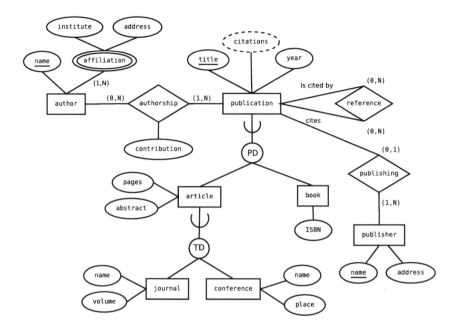

Fig. 1. A citation-enhanced bibliographic database

The size of the expression $\rho(a_1, \ldots, a_n)$ is $n \cdot (n+1)/2$. Furthermore, the regular expression is deterministic, in the sense that its standard automata-theoretic translation is a deterministic automaton. This is relevant since both DTD and XML Schema content models must be deterministic.

As a full example of the mapping, consider the ER schema in Figure 1. It describes a citation-enhanced bibliography, a typical semi-structured data instance. The mapped schema is shown in Figure 2. The XML Schema version is available at ChronoGeoGraph web site [2].

3   Nesting the Structure

Nesting the XML structure has two advantages. The first advantage is the reduction of the number of constraints inserted in the mapped schema and hence of the validation overhead. The second advantage is the decrease of the (expensive) join operations needed to reconstruct the information at query time. Indeed, highly nested XML documents can be better exploited by tree-traversing XML query languages like XPath. We illustrate these points with the following example. Suppose we want to model a one-to-one relationship direction (**dir**) between entities manager (**man**), with attributes **ssn** and **name** and key **ssn**, and department **dep**, with attributes **name** and **address** and key **name**. A manager

```
publication(title, year, citations, reference[0,N],
            authorship[1,N], (article | book)[0,1])
  reference(title)
  authorship(name, contribution)
  article(pages, abstract, (journal | conference))
    journal(name, volume)
    conference(name, place)
  book(ISBM)
publisher(name, address, publishing[1,N])
  publishing(title)
author(name, affiliation[1,N])
  affiliation(institute, address)

KEY(publication.title), KEY(publisher.name)
KEY(author.name), KEY(publishing.title)
KEYREF(reference.title --> publication.title)
KEYREF(authorship.name --> author.name)
KEYREF(publishing.title --> publication.title)
```

Fig. 2. The mapping of the citation-enhanced bibliographic database

directs exactly one department and a department is directed by exactly one manager. In the following, we propose a flat mapping (on the left) and a nested mapping (on the right) of the corresponding ER fragment:

```
man(ssn, name, dir)              man(ssn, name, dir)
  dir(name)                        dir(dep)
dep(name, address)                  dep(name, address)
KEY(man.ssn)                     KEY(man.ssn)
KEY(dep.name)                    KEY(dep.name)
KEY(dir.name)
KEYREF(dep.name  --> dir.name)
KEYREF(dir.name --> dep.name)
```

Notice that nesting saves three constraints over five (one key and two foreign keys). Furthermore, suppose we want to retrieve the address of the department directed by William Strunk. A first version of this query written in XPath and working over the flat schema follows:

```
/dep[name = /man[name = "William Strunk"]/dir/name]/address
```

The query joins, in the outer filter, the name of the current department and the name of the department directed by William Strunk. This amounts to jump from the current node to the tree root. An alternative XPath query tailored for the nested schema is given in the following:

```
/man[name = "William Strunk"]/dir/dep/address
```

This version of the query fluently traverses the tree without jumps. The same happens if the query is written in XQuery. We expect that the second version of the query is processed more efficiently by most XML query processors.

In the mapping proposed in Section 2, nesting is achieved by using specializations and total functional relationships, which are relationships such that one of the participating entities has a participation constraint equal to (1,1). While the nesting structure of specialization is uniquely determined, this is not always the case with the nesting structure induced by total functional relationships. Indeed, it may happen that some entity can be nested in more than one other entity, generating a *nesting confluence*. Moreover, *nesting loops* can occur. Both nesting confluences and nesting loops must be broken to obtain a hierarchical nesting structure. This can be done, however, in different ways. Hence, the problem of finding the *best* nesting structure arises.

In the following, we formalize the nesting problem in graph theory. Let S be an ER schema and $G = (V, E)$ be a directed graph such that the nodes in V are the entities of S that participate in some total functional relationship and $(A, B) \in E$ whenever there is a total functional relationship R relating A and B such that B participates in R with cardinality constraint (1,1). Hence, the direction of the graph edges indicates the entity nesting structure, that is, $(A, B) \in E$ whenever entity A contains entity B. We call G the *nesting graph* of S. A nesting confluence corresponds to a node in the graph with more than one predecessor and a nesting loop is a graph cycle. A spanning forest is a subgraph G' of G such that: (i) G' and G share the same node set; (ii) each node in G' has at most one predecessor; (iii) G' has no cycles. Notice that each spanning forest is a valid nesting solution since it contains neither confluences nor cycles. In general, however, a graph has (exponentially) many spanning forests. We are ready to define the following two nesting problems:

The maximum connectivity nesting problem (MCNP). Given a nesting graph G for an ER schema, find a maximum connectivity spanning forest (MCSF), that is, a spanning forest with the maximum number of edges, or, equivalently, with the minimum number of trees;

The maximum depth nesting problem (MDNP). Given a nesting graph G for an ER schema, find a maximum depth spanning forest (MDSF), that is, a spanning forest with the maximum sum of node depths.

Notice that both problems always admit a solution which is not necessarily unique. The MCNP finds a forest that minimizes the number of schema constraints when the forest is used to nest the entities. Indeed, as shown above, each nesting edge reduces the number of constraints in the mapped schema and hence the larger is the number of edges in the nesting graph, the lower is the number of constraints in the resulting schema. On the other hand, the MDNP finds a forest that minimizes the number of join operations that are necessary to reconstruct the information at query time. Indeed, the deeper is a node in the nesting forest, the larger is the number of nodes belonging to the nesting path containing that node, and the lower is the chance of requiring a join operation in a query involving that entity.

The reader might wonder if a spanning forest with the maximum connectivity is also a spanning forest with the maximum depth. The answer in negative, as shown in the example depicted in Figure 3.

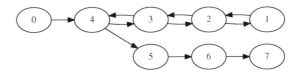

Fig. 3. A nesting graph: a MCSF is obtained by removing edges (1,2), (2,3), and (3,4). It is composed of one tree, 7 edges, and the sum of node depths is 19. A MDSF is the simple path from node 1 to node 7 plus the node 0. It comprises 2 trees, 6 edges, and the sum of node depths is 21. Notice that in this case both solutions are unique.

In the following, we describe an efficient algorithm MCSF that finds a maximum connectivity spanning forest of a nesting graph G. A root node in a directed graph is a node with no incoming edges. Given a node v in a graph G, a reachability tree of v is a directed tree rooted at v containing a path for each node reachable from v in G. Notice that at least one reachability tree exists for any node v, that is, the shortest-path tree from v. The algorithm MCSF works as follows:

1. compute the graph G' of the strongly connected components (SCCs) of G;
2. let $C_1, \ldots, C_k$ be the root nodes in G'. For i from 1 to k, compute the reachability tree $T(C_i)$ rooted at some node in C_i in the graph G and remove all nodes in $T(C_i)$ and all edges involving nodes in $T(C_i)$ from G;
3. output the forest obtained by taking the disjoint union of all trees $T(C_i)$ for i from 1 to k.

Figure 4 illustrates an execution of the sketched algorithm.

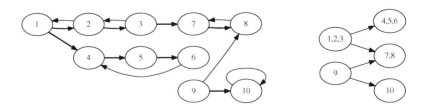

Fig. 4. A nesting graph (left) and its SCC graph (right). Starting the visit from the root component $\{1,2,3\}$, the resulting MCSF is shown in bold. A second MCSF is obtained by starting the visit from the root component $\{9\}$.

Theorem 1. *MCSF computes a maximum connectivity spanning forest of a nesting graph G in time linear in the size of G.*

Proof. As far as the complexity of algorithm MCSF is concerned, the only crucial point is the computation of the SCCs which can be performed in time $\Theta(|V|+|E|)$ exploiting a double depth-first search [5].

As for correctness, we start observing that the MCNP is equivalent to the problem of finding a spanning forest with minimum number of tree roots, since in a graph with n nodes a forest has $n - k$ edges if and only if it has k roots. We proceed by induction on the number of SCCs of G. In the base case, suppose G has one SCC. Then we can build a spanning forest having just one root, since each node reaches all the nodes of G. In this case our algorithm correctly computes a spanning forest having one root.

As for the inductive step, let us assume that we have proved the correctness of our algorithm on graphs having at most $r - 1$ SCCs and let G be a graph with r SCCs. The SCC graph G' of G is an acyclic graph, hence there exists at least one node C in G' without outgoing edges, i.e., C is a SCC of G that does not reach any other SCC. We distinguish two cases: (1) in G' the node C has at least one incoming edge; (2) in G' the node C has no incoming edges. In case (1) a spanning forest of G having minimum number of roots has the same number of roots of a spanning forest of $G \setminus C$ having minimum number of roots. By inductive hypothesis our algorithm is correct on $G \setminus C$, i.e., it determines a spanning forest having the correct number of roots. Moreover, since C has at least one incoming edge in G', we have that C is not used by our algorithm as root node of G'. Hence, our algorithm determines on G a forest having the same number of roots of that determined on $G \setminus C$. This means that our algorithm is correct on G. In case (2), if a spanning forest of $G \setminus C$ having minimum number of roots has k roots, then a spanning forest of G having minimum number of roots has $k + 1$ roots (and vice versa). In this case, since C does not reach and is not reached by other components of G, our algorithm determines one tree rooted at C and works on the remaining components as it works on $G \setminus C$. Hence, since by inductive hypothesis it is correct on $G \setminus C$, it is correct also on G.   □

On the other hand, the MDNP is hard and, unless P = NP, there is no efficient algorithm that solves this problem.

Theorem 2. *The maximum depth nesting problem is NP-complete.*

Proof. (Sketch) We recall that an optimization problem is NP-complete if the (standard) decision problem associated with it is NP-complete [5]. The decision problem associated with our problem consists in deciding whether a graph has a spanning forest of depth k. In the rest of this proof we denote such a problem as DDNP.

Let $G = (V, E)$ be a directed graph and F be a spanning forest of G. Let the depth of F, denoted by S_F, be the sum of depths of nodes in F. We say that a spanning forest is a chain if it contains $|V| - 1$ nodes having one outgoing edge and one leaf. We claim that: (1) $S_F \leq (|V| \cdot (|V|-1))/2$; (2) $S_F = (|V| \cdot (|V|-1))/2$ iff F is a chain. The idea behind the proof is as follows: to maximize the depth of a generic forest the nodes has to be pushed as deep as possible, leading to a chain of nodes whose depth is clearly $(|V| \cdot (|V| - 1))/2$.

It is easy to see that DDNP is in NP: given a spanning forest F, its depth S_F can be computed in polynomial time. We show that DDNP is NP-hard by reducing the Hamiltonian path problem – the problem of deciding whether there exists a path that visits each node of a graph exactly once – to it. The above claim allows us to prove that G has an Hamiltonian path if and only if there G has a spanning forest of depth $(|V| \cdot (|V|-1))/2$. Indeed, if G has an Hamiltonian path H, then H is a spanning forest of G. Moreover, since an Hamiltonian path is a chain, we have $S_H = (|V| \cdot (|V|-1))/2$ (point (2) of the claim). On the other hand, if G has a spanning forest of depth $(|V| \cdot (|V|-1))/2$, then by point (2) of the claim G has a spanning forest which is a chain. A chain is nothing but an Hamiltonian path, hence our graph has an Hamiltonian path.   □

4   Related and Future Work

There is a vast literature about the integration of XML with relational databases – see [6] for a general comparison of XML and relational concepts and for basic kinds of mappings between them. We found, however, that this literature in partly redundant (even when contributions come from the same authors) and the corresponding citation network is quite disconnected. Nevertheless, we identified three main research themes connected to our work. The research theme closest to the present contribution is that of mapping ER conceptual schemas into some XML schema language. We would like to mention a couple of contributions: Kleiner and Lipeck [7] present a mapping from ER with specialization into DTD that preserves information about the structure and as many constraints as possible. The expressiveness limitations of DTD, however, reduce the number of preserved constraints, complicate the mapping, and do not allow a full reversibility of the design. Elmasri et at. [1] design a system for generating user-customized hierarchical views of an ER schema, creating XML Schemas from such views and generating SQL queries to build XML instance documents from data in a relational database with the same conceptual schema. In particular, they describe an algorithm to eliminate graph cycles in the ER diagram subset selected by the user.

A second related research topic is the translation of relational logical schemas into some XML schema language. It is worth noticing, however, that, from the point of view of integrity preservation, converting relational logical schemas is easier than converting ER conceptual schemas. Indeed, as pointed out in Section 2, the relational model allows to specify fewer integrity constraints than the ER model. An informative contribution in this research line is [8], which includes a survey of different techniques of representing many-to-many relationships within XML and gives many references to related works. In particular, Duta et al. [9] propose algorithms for transforming relational schemas to XML Schema considering the following metrics in this order: constraint-preservation, nested structure, compact structure, length of generated XML file, similarity to the relational structure. The authors state that the incorporation of a query metric (as investigated in this paper) in the translation criteria would be desirable.

A third relevant research thread is the development of conceptual models for XML databases. Proposals include suitable extensions of the ER model, e.g., the ERX model [10], models based on UML [11] and ORM [12], and hierarchical models, like ORA-SS [13]. Besides proposing a new conceptual model tailored for XML, most contributions in this research line give automatic procedures to map the conceptual model to some XML schema language. See [14] for a survey comparing many conceptual models for XML including more references.

Future work comprises the investigation of polynomial-time approximation algorithms for the maximum depth nesting problem and the integration of the algorithms in the translation module. Moreover, we intend to test the translation, validation, and query performance on a realistic case study.

References

1. Elmasri, R., Li, Q., Fu, J., Wu, Y.C., Hojabri, B., Ande, S.: Conceptual modeling for customized XML schemas. Data and Knowledge Engineering 54(1), 57–76 (2005)
2. Gubiani, D., Montanari, A.: ChronoGeoGraph: an expressive spatio-temporal conceptual model. In: SEBD, pp. 160–171 (2007), http://dbms.dimi.uniud.it/cgg/
3. Elmasri, R., Navathe, S.B.: Fundamentals of Database Systems, 5th edn. Addison-Wesley, Reading (2007)
4. Liu, C., Vincent, M.W., Liu, J.: Constraint preserving transformation from relational schema to XML Schema. World Wide Web 9(1), 93–110 (2006)
5. Cormen, T.H., Stein, C., Rivest, R.L., Leiserson, C.E.: Introduction to Algorithms. McGraw-Hill Higher Education, New York (2001)
6. Kappel, G., Kapsammer, E., Retschitzegger, W.: Integrating XML and relational database systems. World Wide Web 7(4), 343–384 (2004)
7. Kleiner, C., Lipeck, U.W.: Automatic generation of XML DTDs from conceptual database schemas. In: GI Jahrestagung (1), pp. 396–405 (2001)
8. Link, S., Trinh, T.: Know your limits: Enhanced XML modeling with cardinality constraints. In: ER, pp. 19–30 (2007)
9. Duta, A.C., Barker, K., Alhajj, R.: Conv2XML: Relational schema conversion to XML nested-based schema. In: ICEIS, pp. 210–215 (2004)
10. Psaila, G.: ERX: A conceptual model for XML documents. In: SAC, pp. 898–903 (2000)
11. Combi, C., Oliboni, B.: Conceptual modeling of XML data. In: SAC, pp. 467–473 (2006)
12. Bird, L., Goodchild, A., Halpin, T.A.: Object role modelling and XML-Schema. In: Laender, A.H.F., Liddle, S.W., Storey, V.C. (eds.) ER 2000. LNCS, vol. 1920, pp. 309–322. Springer, Heidelberg (2000)
13. Dobbie, G., Xiaoying, W., Ling, T., Lee, M.: Designing semistructured databases using ORA-SS model. In: WISE (2001)
14. Necasky, M.: Conceptual modeling for XML: A survey. In: DATESO (2006)

Atomicity for XML Databases*

Debmalya Biswas[1], Ashwin Jiwane[2], and Blaise Genest[3]

[1] SAP Research, Vincenz-Priessnitz-Strasse 1, Karlsruhe, Germany
debmalya.biswas@sap.com
[2] Department of Computer Science and Engineering, Indian Institute of Technology,
Mumbai, India
ashwinjiwane@cse.iitb.ac.in
[3] IRISA/CNRS, Campus Universitaire de Beaulieu, Rennes, France
genest@crans.org

Abstract. With more and more data stored into XML databases, there
is a need to provide the same level of failure resilience and robustness that
users have come to expect from relational database systems. In this work,
we discuss strategies to provide the transactional aspect of atomicity
to XML databases. The main contribution of this paper is to propose
a novel approach for performing updates-in-place on XML databases,
with the undo statements stored in the same high level language as the
update statements. Finally, we give experimental results to study the
performance/storage trade-off of the updates-in-place strategy (based on
our undo proposal) against the deferred updates strategy to providing
atomicity.

1 Introduction

With more and more data stored into XML databases, there is a need to provide
the same level of failure resilience and robustness that users have come to expect
from relational database systems. A key ingredient to providing such guarantees
is the notion of transactions. Transactions have been around for the last 30 years
leading to their stable and efficient implementations in most current commercial
relational database systems. Unfortunately, (to the best of our knowledge) there
still does not exist a transactional implementation for XML databases. Here,
we are talking about native XML databases [1], and not relational databases
extended to store XML data [2] that in turn rely on the transactional features of
the underlying relational database. The need to be able to update XML however
has been widely recognized and addressed by both industry and researchers with
the recent publication of the W3C standard XQuery Update Facility 1.0 [3].
Researchers have also started exploring efficient locking protocols for XML [4,5].
In this work, we discuss strategies to provide a transactional implementation for
XML databases based on [1,3], with specific focus on the atomicity aspect.

* This work was partially done while the first and second authors were at
IRISA/INRIA, France. This work is supported by the CREATE ACTIVEDOC,
DOCFLOW and SecureSCM projects.

Z. Bellahsène et al. (Eds.): XSym 2009, LNCS 5679, pp. 180–187, 2009.
© Springer-Verlag Berlin Heidelberg 2009

Transactions [6] provide an abstraction for a group of operations having the following properties: A (Atomicity), C (Consistency), I (Isolation), D (Durability). Atomicity refers to the property that either all the operations in a transaction are executed, or none. There are usually two strategies for performing updates in an atomic fashion:

- Deferred Updates (UD): In this strategy, each transaction T has a corresponding private workspace W_T. For each update operation of T, a copy of the data item is created in W_T, and the update applied on the local copy. Upon commit of T, the current values of the data items in W_T need to be reflected atomically to the actual database. As obvious, abortion is very simple and can be achieved by simply purging W_T.
- Updates-in-Place (UIP): Here, transactional updates are applied as and when they occur. However, the disadvantage of UIP is the added complexity in the abortion process. To provide atomicity in such a scenario, the "before images" of all updated data items is usually maintained in a log, which can then be used to undo the updates if required.

In this work, we study both approaches for providing atomicity to XML databases. The main contribution of this paper is to propose a novel approach for providing UIP on XML databases. Rather than maintaining the "before images" *data*, our proposal stores undo *operations* written in the high level query/update language. Basically, for each update, it generates the undo statements dynamically (at run-time) that can be used to undo the effects of that update. The transformation rules to dynamically generate the undo operations are presented in Section 3.

The performance trade-off between UD and UIP depends upon the additional overhead of creating and writing copies in UD, and depends upon the time taken to create the undo operations and perform them in the event of a failure in UIP. The amount of storage required to store the undo data is also an area of concern for current database implementations. Long running transactions performing large number of updates often have to be aborted due to insufficient disk space (e.g., the "ORA: Snapshot Too Old" error). Thus, the undo storage requirements of both strategies are important. In our case, the UD strategy in particular leads to storage/performance trade-offs. Recall that for each update, the UD strategy first creates copies of the data items (if they do not already exist in its private workspace), and then performs the updates on the copies. The additional create step per update can clearly be avoided if we create a copy of the whole XML document for the first update on a document, then any subsequent updates on nodes of that document would not need the additional create step. Obviously, this is not efficient from a storage perspective as it would lead to storage redundancy unless all the nodes of a document are updated (as part of one transaction).

We have implemented both the UD and UIP strategies for XML databases, and provide experimental results with respect to the performance/storage trade-off between the two strategies for the issues identified above in Section 4.

Note that this approach of using undo statements to preserve atomicity is in line with using compensation [9] to semantically undo the effects of a transaction. Here, it helps to recall that compensation is not equivalent to the traditional database "undo", rather it is another forward operation that moves the system to an acceptable state on failure. While compensation mechanisms have been accepted and are available in high level languages (e.g., compensation handlers in BPEL [8]), they have not been studied for database updates (at least, not explicitly). Implicitly, when we say that a bank withdrawal can be compensated by a deposit operation, it corresponds to the effects of an update SQL being compensated by another update SQL on the accounts table.

2   XML Update Syntax

In this section, we give a brief introduction to the XQuery Update Facility (XUpdate) [3] for performing XML updates. XUpdate adds five new kinds of expressions, called insert, delete, replace, rename, and transform expressions, to the basic XQuery model [7]. For simplicity, we only focus on the insert, delete and replace expressions, also referred to as the update expressions in general. The main differences between XUpdate and SQL, apart from the hierarchical nature of XML, arises from the significance of a node's location in the XML document (XML documents are basically ordered trees.). For example, let us consider the syntax of the XUpdate insert expression:

InsertExpr ::= "insert" ("node" | "nodes") SourceExpr TargetChoice TargetExpr
TargetChoice ::= (("as" ("first" | "last"))? "into") | "after" | "before"

An insert expression is an updating expression that inserts copies of zero or more source nodes (given by *SourceExpr*) into a designated position with respect to a target node (given by *TargetExpr*). The relative insert position is given by *TargetChoice* having the following semantics: If *before* (or *after*) is specified, then the inserted nodes become the preceding (or following) siblings of the target node. If *as first into* (or *as last into*) is specified, then the inserted nodes become the first (or last) children of the target node. For the detailed specifications of *InsertExpr*, *DeleteExpr* and *ReplaceExpr*, the interested reader is referred to [3].

The evaluation of an update expression results in a pending update list of update primitives, which represents node state changes that have not yet been applied. For example, if *as first into* is specified in the given insert expression, then the pending update list would consist of update primitives of the form:

insertIntoAsFirst($target, $clist)

The effects of the above primitive can be interpreted as: Insert *$clist* nodes as the leftmost children of the *$target* node. Note that the node operands of an update primitive are represented by their node identifiers.

The pending update lists generated by evaluation of different update expressions can be merged using the primitive *mergeUpdates*. Update primitives are held on pending update lists until they are made effective by an *applyUpdates* primitive.

We consider a pending update list consisting of a sequence of update (insert/delete/replace) primitives as a transactional unit, with the *applyUpdates* primitive acting as the corresponding Commit operation. The objective then is to ensure that the update primitives in a pending list execute as an atomic unit, i.e. either all the update primitives are applied or none. In the next section, we show how to generate the corresponding undo primitive that can be used to undo the effects of an update primitive in the event of a failure.

3   XML Undo Primitives

The underlying intuition of undo primitives is that the effects of an insert (delete) primitive can be canceled by the subsequent execution of a delete (insert) primitive. For example, for the update primitive:

insertBefore (t, {n_1, n_2, n_3})

its undo primitive(s) would be:

delete(n_1)
delete(n_2)
delete(n_3)

where t, n_1, n_2, n_3 refer to node identifiers. Note that an update primitive may lead to more than one undo primitive, and vice versa. Basically, for a given pending update list, as its update primitives are processed one by one, we assume that the same processor can use our mechanism to simultaneously generate (and store) the respective undo primitives.

We give the undo generation rules of the different update primitives in the sequel.

3.1   Insert

We start with the insert primitive.

insertBefore ($target as node(), $content as node()+)

Let $content = \{n_1, \cdots, n_m\}$. Then, its undo primitives are:

delete(n_m), $\cdots$, delete(n_1)

The undo primitives for the other insert primitives: *insertAfter*, *insertInto*, *insertIntoAsFirst* and *insertIntoAsLast*, can be generated analogously.

3.2 Delete

The undo primitive generation of the delete primitive is slightly more complicated. This is because of the lack of position specifier in the delete primitive:

delete($target as node())

That is, we do not know the position of the node to be deleted in the XML
document, and consequently do not know where to re-insert it (if required) to
undo the effects of the delete. If we assume some correlation between the node
identifier and its position in the document, then we can infer the position of the
node from its identifier. However, [3] does not impose any such restrictions on
the node identifier generation scheme, and neither do we assume the same here.
In the absence of such correlation, we first need to get the position details of the
node to be deleted before generating the corresponding undo primitive. There are
of course several ways of determining the relative position of the $target node in
the document, and the idea is to minimize the number of function calls needed.
An approach to generate the undo primitive based on the relative position of a
sibling node would be as follows:

$sibling = getNextSibling($target)
insertBefore ($sibling, $target)

The variant of the above approach based on the preceding sibling would be
as follows:

$sibling = getPreceedingSibling($target)
insertAfter ($sibling, $target)

While the above approaches require only a single additional function call, they
would not work if the $target node does not have any siblings (is the only child).
Then, we need to get the position of $target relative to its parent node.

$parent = getParent($target)
insertInto ($parent, $target)

Note that in the worst case (no siblings), the above approach leads to three
function calls, i.e. *getNextSibling()*, followed by *getPreceedingSibling()*, and finally *getParent()*. An alternate approach that also needs two additional function
calls would be as follows:

$parent = getParent($target)
$listChildren = getChildren($parent)
Let i be the position of *$target* in *$listChildren*, and n_i refer to the node identifier
at position i in *$listChildren*.
if ($i = 0$)
    then *insertInto ($parent, $target)*
    else *insertAfter (n_{i-1}, $target)*

Clearly, which approach is better depends on the underlying XML data characteristics, and would be difficult to predict in advance.

We can offset the time taken by the additional functional calls by doing some pre-processing. Note that on *applyUpdates()*, the update primitives in a pending update list are processed in the following order [3]: insert, followed by replace, finally followed by delete primitives. As the delete primitives are processed at the end, the positions of the delete primitives can be pre-determined in parallel while processing of the preceding insert and replace primitives.

3.3   Replace

The replace primitive is given below:

replaceNode ($target as node(), $replacement as node()*)

The semantics of the replace primitive is to replace the *$target* node by the *$replacement* node(s). A replace primitive can be implemented as a combination of insert and delete primitives as follows:

replaceNode ($target, $replacement)
    $\Leftrightarrow$
insertAfter ($target, $replacement), delete ($target)

The undo primitives for replace can then be constructed as follows: Let $replacement = \{n_1, \cdots, n_m\}$.

insertAfter (n_m, $target)
delete(n_m), $\cdots$, delete(n_1)

The first *insertAfter* undo primitive relies on the semantics that insertion of a list of nodes preserves their original source order, i.e. n_2 is the right sibling of n_1, $\cdots$, n_m is the rightmost sibling node to be inserted. Note that here we do not need the additional function calls to generate the undo primitive of delete, as is needed for standalone delete primitives (Section 3.2).

4   Experimental Analysis: Performance and Storage

The experiments were performed on the Active XML (AXML) [10] infrastructure. AXML provides a P2P based distributed XML repository. The XML repositories in AXML are implemented based on the *Exists* XML database. *Exists* currently does not provide any transactional features, at least not at the user level. We have extended *Exists* to implement the transactional aspects discussed in this paper.

As mentioned earlier, the main factors affecting the performance trade-off between UD and UIP (implemented based on our proposal) can be summarized as follows: the additional overhead of creating and writing copies in UD versus the additional time needed to perform the undo operations in the event of a failure. Fig. 1 gives some comparison results.

Clearly, if there are no failures, UIP is the fastest (no undoing is required, while the time taken to create the undo operations is negligible). In the event

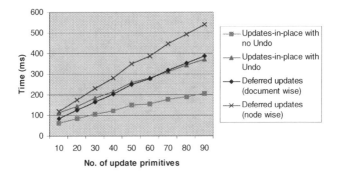

Fig. 1. Execution Time vs. Number of update primitives per transaction

%	Execution Time (in ms)		% Increase in
	Document	Node	Execution Time
10	47	74	57.44680851
20	60	93	55
30	80	113	41.25
40	93	133	43.01075269
50	108	152	40.74074074
60	124	169	36.29032258
70	135	189	40
80	152	213	40.13157895
90	165	221	33.93939394

Fig. 2. Performance/storage trade-off

of failures, undos are required. When the number of operations is small, (document wise) UD fares slightly better. However, as the number of operations in a transaction increases, this advantage seems to fade away. That is, in mean value considering a reasonable number of failures, we expect UIP based on our proposal to perform better than UD, and in particular for large transactions.

Fig. 2 shows the storage/performance trade-off between creating copies of the whole document against that of the specific affected nodes at a time in UD. If 10% of the nodes of a document are updated, then creating and updating copies node wise takes 57% more time than doing it document wise. On the other hand, the storage space is ten times less than that required by the document wise strategy. On the other hand, if 90% of the the document nodes are updated, then the nodewise strategy only needs 33% more time than the document wise strategy, while saving only 10% of storage space. While it is expected for the overall execution time of the node wise strategy to increase as the percentage of document nodes updated increases, the interesting result is that the percentage difference in execution time actually decreases (from 57% to 34%) as the percentage node updation increases (from 10% to 90%). We infer that this decrease is because as the percentage of nodes updated in a document increases, the probability of the same nodes being updated more than once also increases (for whom, the overhead copy creation time is saved). Compared to UIP based on our proposal, the storage space needed for UIP is very close to that needed for node-wise UD strategy. That is, it seems that UIP is better both in terms of storage space and execution time. The only drawback is obviously its more complex implementation.

5   Conclusions

Transactional implementation for XML databases are still in their infancy, with the recent release of the W3C specification to perform XML updates. With updates, comes the natural requirement to be failure resilient, and hence transactions. In this work, we considered the two main approaches for performing updates in an atomic fashion: UD and UIP. We proposed a novel approach for providing UIP on XML data, where the undo data required to perform rollback in the event of a failure/abort is stored in the same high level language as the update statements. This allows for a lightweight implementation (without going into the disk internal details) while providing comparable performance, as verified by experimental results.

References

1. Open Source Native XML Database, http://exist.sourceforge.net/
2. XML Database Benchmark: Transaction Processing over XML (TPoX), http://tpox.sourceforge.net/
3. XQuery Update Facility 1.0 Specification, http://www.w3.org/TR/xquery-update-10/
4. Dekeyser, S., Hidders, J., Paredaens, J.: A Transactional Model for XML Databases. World Wide Web 7(1), 29–57 (2002)
5. Haustein, M.P., Härder, T.: A Transactional Model for XML Databases. J. Data Knowledge Engineering 61(3), 500–523 (2007)
6. Weikum, G., Vossen, G.: Transactional Information Systems: Theory, Algorithms, and the Practice of Concurrency Control. Morgan Kaufmann Publishers, San Francisco (2001)
7. XQuery 1.0: An XML Query Language Specification, http://www.w3.org/TR/xquery/
8. Business Process Execution Language for Web Services Specification, http://www.ibm.com/developerworks/library/specification/ws-bpel/
9. Biswas, D.: Compensation in the World of Web Services Composition. In: Cardoso, J., Sheth, A.P. (eds.) SWSWPC 2004. LNCS, vol. 3387, pp. 69–80. Springer, Heidelberg (2005)
10. Active XML, http://activexml.net

Author Index